THE LITTLE BOOK
of planet
earth

THE LITTLE BOOK

of planet earth

ROLF MEISSNER

DISCARD

C

COPERNICUS BOOKS

An Imprint of Springer-Verlag

Originally published as *Geschichte der Erde*,
© 1999 Verlag C. H. Beck oHG, München, Germany.

© 2002 Springer-Verlag New York, Inc.

Published in the United States by Copernicus Books,
an imprint of Springer-Verlag New York, Inc.
A member of BertelsmannSpringer Science+Business Media GmbH

Copernicus Books
37 East 7th Street
New York, NY 10003
www.copernicusbooks.com

Illustration: The BRM with Jordan Rosenblum

Library of Congress Cataloging-in-Publication Data
Meissner, Rolf.
    [Geschichte der Erde. English]
    The little book of planet Earth / Rolf Meissner.
       p. cm.
    ISBN 0-387-95258-6 (acid-free paper)
    1. Earth. I. Title.
    QB631.M4513   2002
    550—dc21   2001060201

Manufactured in the United States of America.
Printed on acid-free paper.
Translated by Rolf Meissner.

9 8 7 6 5 4 3 2 1

ISBN 0-387-95258-6   SPIN 10796085

The most beautiful and most powerful emotion we can experience is the sensation of the mystical.

—Albert Einstein

Preface

According to Greek mythology, the Earth, *Gaia*, rose out of the Chaos, an imaginative concept not too far removed from the tenets of modern cosmology. Many astronomers believe that several billion years after the Big Bang, rotating clouds of dust and gas in one of the spiral arms of our Milky Way galaxy became increasingly dense. Near the maximum of density, the power of gravity drove most of this rotating debris to the center and formed our Sun. The remnants of this process, about 0.01 percent of the mass in our Solar System, created the planets. Of these planets, the third from the Sun is our Earth. A very fortunate combination of size and proximity to the Sun permitted the creation of water on our planet—the basis of all life.

For several thousand years, man has wondered about the Earth's shape, its position in the Solar System, and its composition. Our knowledge of the Earth's interior, evolution, and dynamics has been growing dramatically since the second half of the twentieth century. The classic saying of Heraclitus (500 years B.C.), "*Panta rhei*," "Everything flows," definitely applies to our planet. Today, we know that the surface and the interior of Earth are indeed mobile, though slowly so. The velocities may be only millimeters or centimeters per year, but over millions of years the movement adds up to thousands of kilometers.

However, until a few decades ago, geologists were divided into two camps over the issue of the Earth's mobility. At the beginning of the twentieth century, Alfred Wegener, the first major proponent of the mobile view of the Earth, combined many observations of geologists, geophysicists, climatologists, and biologists to show how our planet's continents have moved over the last 200 million years. He was the great forerunner of the concept of plate tectonics, which was established in the 1960s—30 years after his death—which led to the conceptual integration of geology and geophysics. Plate tectonics influenced all of the natural sciences, dominated research activities (especially in the conjoined fields of geology and geophysics), and accounted for a huge number of observations from all over the world. It explained the movement of oceanic and continental plates, the formation of mountain ranges, depressions and rifts, the origin of volcanoes and earthquakes, and the formation of deposits of oil, coal, and ores. Even the origins of life and its development are closely connected to the Earth's geologic development with its strong climatic variations.

This book presents the current state of research on our planet's evolution and structure. It seeks to elucidate the complex interconnections of contemporary concepts and ideas in earth sciences, which are the basis for a better understanding of present-day developments on Earth.

Acknowledgments

This book emerged out of the great research and teaching environment at the Institute of Geophysics of the Christian Albrechts University in Kiel, Germany. Friendly and exciting discussions with colleagues, students, and nonspecialists gave me the idea to write a general and understandable overview of the planet Earth. It should initiate interest and excitement about both the basic ideas and the fast development of earth sciences.

I wish to express my sincere appreciation to editor Dr. Stephan Meyer of C. H. Beck, who helped develop the original German edition of *The Little Book of Planet Earth*. Thanks also to editor Ernest Scott and to Mareike Paessler, Anna Painter, and Paul Farrell of Copernicus Books for their persistent and patient efforts to streamline the English translation, to improve the language, and to attack me with fundamental and apparently simple questions. I also wish to express my thanks to Gudrun Reim in Kiel and Jordan Rosenblum and Andrew Kuo in New York for creating and improving the artwork.

Publisher's Note

A glossary of geological terms can be found at
www.copernicusbooks.com.

The Roots of Earth Sciences

Because scientific inquires and discoveries are and have always been affected by history, a retrospective view of the historical development of the earth sciences seems a good place to begin our discussion of the planet Earth. While this historical survey is necessarily brief—this is, after all, a *Little Book*—it will establish the foundations of modern earth science and introduce important thinkers and scholarly developments.

Classical Scientific Thought

Prior to and even after 600 B.C., most ancient Greeks used mythology to explain the natural world. Then, around 630 B.C., an important intellectual and methodological shift began to occur. Thales, a Greek and the first recorded Western philosopher, began to use a different kind of reasoning as he coupled his own speculations with observations and mathematics. Among other things,

Thales claimed that all of the Earth's matter was made of water and, furthermore, that the Earth was a sphere and not a disk, as was widely believed at the time. About 50 years later, Pythagoras, another Greek philosopher, claimed that the Sun and the Moon also were spheres, and that Mercury and Venus rotated around the Sun. Controversial at the time, this notion of a heliocentric (Sun-centered) system foreshadowed the view of the world that would be established by Copernicus almost 2000 years later.

Around 450 B.C., Herodotus, who was not only a famous Greek historian but may also be considered one of the first sedimentary geologists, observed that the Nile slowly enlarged its delta, and that this deposition process took an extremely long time. About 100 years later, the great philosopher Aristotle wrote some 400 treatises, many of them on physics and biology. Around the year 200 B.C., Eratosthenes at Alexandria's Greek University calculated the exact circumference of the Earth by observing the length of the Sun's shade in two different locations at fixed times. In A.D. 150, Ptolemy, considered the last great classical astronomer, returned to the geocentric view of the world. He argued that the Earth had to be at the center of the Universe, and that the Sun, Moon, and stars revolved around it in very complex "epicycles." After Ptolemy, the scientific knowledge that the Greeks had developed was largely forgotten, and Roman pragmatism limited further developments in the natural sciences.

The Copernican Revolution

Near the end of the Middle Ages, the natural sciences experienced a revival, and a number of discoveries were made that have had profound effects on our understanding of the Universe and our planet. This new era of scientific discovery was initiated by Nicolaus Copernicus in the sixteenth century. In his most famous work, *De revolutionibus orbium coelestium*, which was published shortly before his death in 1543, he outlined the heliocentric system in which the Earth and the visible planets revolve around the Sun. In the early 1600s, Johannes Kepler, a German astronomer who was greatly influenced by Copernican teachings, formulated laws of the planets' movements around the Sun based on very exact observations by his contemporary and teacher Tycho Brahe. About the same time, Galileo Galilei, the great Italian physicist, observed the orbiting planets and four of Jupiter's moons with a simple, self-made telescope. These and other observations convinced Galileo that the Copernican view was true. As a result of these convictions, he had to defend his theories against ecclesiastical authorities in Rome who believed that Copernicus's teachings were dangerous to the faith. Under threats of torture, at an age of 70 years, Galileo renounced his opinion. Ultimately, it took the Church more than 300 years to tolerate the Copernican view of our Solar System. Over 50 years after Galileo's trial, another of the science's most famous sons, Sir Isaac Newton, published his most renowned work, *Philosophiae naturalis principia mathematica*. In it, he developed a substantial physics-based foundation to Kepler's laws and outlined his own theories of motion and gravity. Looking back at the great achievements from Copernicus to Newton, we can see a true revolution in scientific thinking that most convincingly marked the transition from the Middle Ages to modern times. No longer were the study of old philosophers, like Aristotle, or a literal citation of biblical text considered the ultimate goal of research. Instead, exact physical observations and rational deduction opened up the time of enlightenment and a new view of the world.

From Physics and Philosophy to Geology

Improved telescopic observations toward the end of the eighteenth century convinced researchers that our Solar System was only part of a much larger unit of stars, called the Milky Way galaxy. Although both René Descartes and Emanuel Swedenborg had attempted to describe stellar nebulae and their role in the Universe, only the German philosopher Immanuel Kant combined Swedenborg's ideas and Newton's physics to create the first substantial nebula theory. However, Kant's explanation, published in 1755, contained a number of physical problems. In 1796, a French scientist named Pierre-Simon Laplace expanded Kant's ideas, proposing that our Solar System originated as a large, swirling nebula that slowly cooled and contracted into its present form. At the beginning of the nineteenth century, Carl Friedrich Gauss of Göttingen, Germany, combined magnetic observations with new mathematical tools to prove that the Earth's magnetic field originated in its deep interior. This important break-through could explain a great number of observations, from polar lights and the poles themselves to the functioning of compasses. No mystical magnetic mountains were needed anymore. One hundred years after Gauss's deductions, several branches of geophysics, including paleomagnetism and archeomagnetism, opened completely new fields in geophysics and geology.

New physical, chemical, and seismological data were collected toward the end of the nineteenth century, but an exponential increase in knowledge of the structure and interior of our planet started only in the twentieth century. Seismology took the lead and was able to decipher many of the mysteries of the Earth, such as the existence and the shape of the metallic core, the huge silicate mantle, and the thin crust. From the sporadic observations of earthquakes with huge (but primitive) seismometers to the development and installation of modern digital seismometer arrays all over the world, one century witnessed great change.

Geological knowledge developed more or less parallel to the physical discoveries. Around the mid-sixteenth century, scholars began to question the biblical conclusion about the age of the Earth. In 1556, Georgius Agricola used his knowledge of mining and metallurgy to explore geologic strata and thermal gradients. More importantly, Agricola began classifying minerals more systematically and is credited with first applying chronology to the discussion of rock formations. Then, in the eighteenth century, both Abraham Gottlob Werner and James Hutton helped lay the basis for modern geology, developing systems for describing minerals, rocks, and strata that are still used today. Werner, a rather dogmatic *neptunist* (from the Roman god Neptune, ruler of all waters), argued that nearly all rocks originated from a series of chemical precipitations and depositions formed by the water that covered the Earth. He was partly right, at least regarding the origin of sedimentary rocks.

At the beginning of the nineteenth century, Nicholas Desmarest, a French government employee, argued that many rocks had to be of igneous origin, namely the basalts in France's *Massif Central.* He and his followers, among them Hutton who was the most prominent researcher, were called the *vulcanists,* or *plutonists* (after the Roman god Pluto, the smith and ruler of the deep fire). They were, correctly, convinced that deep intrusions and many igneous rocks had their ultimate origin at great depth and were similar to volcanic rocks. During the nineteenth century, the controversy between plutonists and vulcanists on one side and neptunists on the other quickly faded. A new research technique, the identification of rocks and the correlation of strata by fossils, laid the basis for modern geology.

At this time, Alexander von Humboldt, a German explorer, observed rocks, strata, and biological species around the world and found interesting correlations. Von Humboldt can be considered a forerunner of the great pioneer of the biological and geological sciences in the second half of the nineteenth century, Charles Darwin. Darwin's travels led him to many parts of the world, and, in a grand synthesis of all his observations, he wrote *On the Origin of Species* and put forth his renowned concept of natural selection. Darwin developed a new understanding of the evolution of life, based on careful investigations of fossils in rocks. His remarkable intellectual leaps laid the foundation of paleontology, linked biology and geology, and greatly advanced studies of the development of animals and humans.

The Age of the Earth

As the astronomical and physical explanations of mysterious celestial and terrestrial phenomena became more known in European universities at the beginning of the eighteenth century, some animosities between theology and science began to develop. Both astronomers and neptunists postulated an age of at least 75,000 years for the origin of the Earth, while some interpreters of the Bible insisted on an age of 4000 to 14,000 years. In fact, around 1654, the Irish Archbishop James Ussher claimed that the Earth was created on October 26, 4004 B.C.

A major break-through occurred in 1860, when the eminent physicist William Thomson Kelvin calculated an age of 25 million years. Kelvin based this estimate of the Earth's age on internal heat flow, which he considered a remnant of the molten stage of the Earth. Later, he adjusted his age calculation multiple times, upward to a maximum of 100 million years. Charles Darwin, who based his estimate on the development of several species, dated the Earth at even 300 million years, but later—under attack from many geologists at various conferences in Great Britain—reduced the number to 100 million years. Most nineteenth-century geologists, observing the slow sedimentation rates in rifts and depressions, estimated an age of 80 to 100 million years.

Then, in 1896, Antoine Henri Becquerel discovered the phenomenon of radioactivity, the geological significance of which was recognized by Pierre and Marie Curie as a new and important source of the Earth's heat. The Curies had first studied radium salts and found a steady flow of heat. Later, they found more heat-releasing radioactive elements. As a result, it was recognized that the Earth's heat was not only coming from a "molten stage" but also from the radioactive elements, mainly uranium, thorium, and potassium.

Shortly thereafter, Ernest Rutherford, a British physicist, became the leading researcher of radioactivity and atomic structure. He studied various radioactive elements and calculated the origin of rocks based on the decay of radioactive elements. The age of these rocks and, consequently, the age of the Earth were surprisingly large. In 1904, he seems to have even convinced old Lord

Kelvin. In fact, in a talk he gave where Kelvin was in the audience, he said that his incorrect age estimates were compatible with Kelvin's early remark, "provided that no new sources of heat were discovered."[1] From this research, new estimates of the age of the Earth started with 500 million years, but continuous improvement in scientific techniques steadily increased that number. Finally, in the middle of the twentieth century, a value of 4600 million years was established, and today the age of the Earth is no longer the subject of major controversy.

Through the long process of theoretical and experimental research, the physical age determination of minerals, rocks, and strata developed into one of the most intensive and successful fields of geology. Today, so-called *isotope geology* is applied to dating rocks on the Earth and from the Moon and is fully integrated into general geology. It stands as an example for the fact that only a combination of physical and geological reasoning makes up the framework of earth sciences.

1 Anthony Hallam, *Great Geological Controversies* (New York: Oxford University Press, 2nd ed. 1990), 101.

chapter 2

The Earth in the Context of Our Solar System

Even though alternate theories have been proposed, the Kant-Laplace theory is still the basis for most of the contemporary theories that attempt to explain the Earth's origins. Using their explanation as an intellectual springboard, scientists studying astrophysics, astronomy, cosmochemistry, and physics have developed a wealth of new methods, including refined spectroscopic methods with various wavelengths, from ultra-short X-rays to very long radio waves. Today, our Sun, the Solar System, and the whole Milky Way galaxy have been extensively explored, and their chemical abundance has been determined. The abundance of metals, e. g. iron, increases toward the center of our Galaxy and decreases toward the outside. At the end of the twentieth century, computer-enhanced and satellite-based telescopes improved our knowledge of galaxies billions of light years away.

The Origins of the Solar System

Several billion years before our Solar System formed, a huge star, a so-called supernova at least 10 times larger than our Sun, collapsed, exploded, and distributed its remnants throughout the galactic neighborhood. The collapse generated a massive cloud of dust and gas that consisted of 99 percent hydrogen and helium, and of only 1 percent heavier, metallic elements. These materials provided the foundation for our planetary system, the Solar System. The cloud of dust and gas circled around the center of our Milky Way galaxy on a spiral arm. It took about 250 million years to make one complete rotation, which is the same speed as that of our present Solar System circling around the center of the Milky Way.

After about 3 billion years, the cloud of dust and gas collapsed. It seems that a critical density was surpassed, a process we observe relatively often in dense clouds in our Galaxy. Possibly, another supernova sent out shock waves that initiated a process of contraction. But whatever the case, the slowly rotating cloud greatly increased its rate of rotation as it collapsed, just as an ice skater gains speed when he draws his arms and legs closer to his body as he spins. Slowly, the cloud material assumed the shape of a disk, as an equilibrium between gravity and centrifugal forces was established. More than 99 percent of the mass accumulated at the center of the disk, and our Sun was formed. The accumulating mass, with enormous pressure and temperature, initiated nuclear fusion. Hydrogen was transformed to helium, and the energy of that fusion process radiated light and heat into the surrounding space. There, remnants of the condensing material formed the planets—both the terrestrial (Earth-like) planets, which are rather near to the Sun, and the larger, mainly gaseous, planets farther away.

EVENT	YEARS BEFORE PRESENT
The Big Bang: Origin of our Universe.	15 ± 1 billion
Formation of the galaxies.	11 ± 1 billion
Formation of quasars.	$10 - 1$ billion
Our dust and gas cloud forms.	$9 - 6$ billion
The cloud begins to collapse.	≈ 4.7 billion
Our Solar System evolves.	≈ 4.6 billion
Formative phase: Heavy meteorite bombardment.	$4.6 - 3.9$ billion
Archean age: Start of tectonic and biological developments. First bacteria and algae in water, beginning of photosynthesis.	$3.9 - 2.5$ billion
Proterozoic age: From single cells to multiple cells. Carbon dioxide level decreases, oxygen level increases.	$2.5 - 0.6$ billion
Palaeozoic era: Skeletons develop, organisms conquer the land, reptiles evolve.	$600 - 230$ million
Mesozoic era: Extinctions and development of new species (birds, dinosaurs, first mammals).	$230 - 65$ million
Cenozoic era/Tertiary period: Extinction and development of new species. Mammals diversify and multiply.	$65 - 1$ million
Hominids and apes.	5.5 million
Homo neanderthalensis and *Homo sapiens.*	0.1 million

TABLE 2.1 Important Events in the History of the Earth

The hydrogen and helium could not remain on the relatively small inner planets because they were too light, and the planets' gravity was too low. A young star like the early Sun emits a strong *solar wind* into its surrounding. Even today, the solar wind is observed as a steady particle radiation from the hot Sun that blows at a rate of 200 to 1000 kilometers a second. In the beginning, however, it would certainly have been much stronger. The solar wind blew the light elements further out, where they accumulated around the cores of Jupiter, Uranus, and Saturn. The largest planet, Jupiter, could have also started fusing hydrogen if it had been slightly larger. If this had happened, our Solar System would now have a binary, or double, star—a phenomenon we often observe in the Universe, where 50 percent of all planetary systems have double or even triple stars.

The Elements of the Solar System

In spite of their common origin, the Sun and its planets are chemically very different from one another. For example, the Sun mainly consists of light elements, such as hydrogen and helium. Eventually, all of the hydrogen will be burnt down to helium, a process that will take about 10 billion years, 4.6 billion of which are past. The steadily increasing radiation of the Sun will make conditions on our planet rather unpleasant. After the next billion years, the oceans will dry out, and Earth will become uninhabitable for higher life forms. It will take about 5 billion years until the Sun has burnt all its hydrogen. When that happens, the Sun will transform into a *red giant* and later into a *white dwarf* (as will all stars of similar size).

Figure 2.1 shows the relative abundance of chemical elements in the Solar System and our Galaxy. Roughly speaking, the lighter the element, the greater its abundance, although the "saw tooth" nature of the curve reflects the fact that some elements are unstable and thus less abundant. The horizontal (x) axis in Figure 2.1 shows the nucleon number (the total number of protons and neutrons) in a typical atom of each element. The vertical (y) axis of the figure is logarithmic; hydrogen (a) and helium (b) are about 1000 times more abundant than are elements like oxygen, magnesium, and silicon (e, f, and g).

The relatively strong abundance of iron is also interesting. The fusion processes in the Sun can only create helium. A supernova more than 10 times the size of our Sun, however, is capable of creating the heavy elements, such as iron. The collapse of a giant supernova in our galactic neighborhood around 9 to 6 billion years ago must have triggered the formation of the elements that made up the solar nebula. In our Galaxy, the abundance of iron increases toward the center, while farther outside the quantity of iron decreases and may hamper the formation of any solid planets. Because of its strong presence in the Universe, Earth, Mercury, and Mars, and presumably all other planets of our Solar System, have metallic cores, mainly consisting of iron.

Another heritage of the rotating cloud of dust and gas from the time of the actual collapse is the similar direction of rotation of all planets around the Sun. All move on disk-like planes with elliptical, for the most part near-circular orbits. (The Earth's orbit is called the *ecliptic.*)

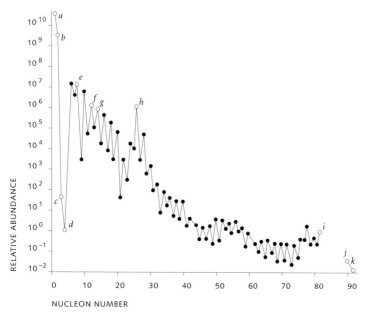

a HYDROGEN
b HELIUM
c LITHIUM
d BERYLLIUM
e OXYGEN
f MAGNESIUM
g SILICON
h IRON
i LEAD
j THORIUM
k URANIUM

FIGURE 2.1 A graphical distribution showing the abundance of the elements in our Solar System based on spectroscopic measurements and chemical analyses of meteorites. Hydrogen (a) is about 1000 times more abundant than oxygen (e).

The Planets Circling the Sun

Figure 2.2 shows the Solar System and the orbits of its planets. The orbits are believed to have not changed much since the formation of our planetary system and conform to the laws that were first developed by Kepler and later physically explained by Newton. As mentioned before, the ages of the Moon, some planets, and various meteorites have been determined by radioactive dating techniques. Their ages—all around 4.6 billion years—is also considered the best estimate for the age of the whole Solar System. Radioactive age determination is based on the known decay rates of certain isotopes with a half-life of millions to billions of years, as will be explained in Chapter 9.

Figure 2.3 shows the tremendous difference in size between the Sun and the various planets. Here, we will limit our description to the inner planets, which are divided from the outer planets by the Asteroid Belt between Mars and Jupiter. This region of the Solar System is the home of most meteorites, some of which have highly elliptical orbits that threaten the Earth.

A comparison of the four inner, or terrestrial, planets provides a surprising picture of how, out of the same chemical material, completely different forms of surfaces and atmospheres developed. Comparative planetology, which is conducted by means of spacecraft carrying out various geophysical, geochemical, and optical investigations, has revealed the nature and the evolution of the planets. The absolute highlight of these efforts was certainly the Apollo astronauts' Moon landing. On the Moon, the astronauts measured moonquakes by means of seismometers and observed the impacts of meteorites as well as the mineralogical,

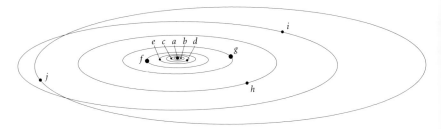

a   SUN
b   MERCURY
c   VENUS
d   EARTH
e   MARS
f   JUPITER
g   SATURN
h   URANUS
i   NEPTUNE
j   PLUTO

FIGURE 2.2 The planets as well as the moons
of our Solar System rotate in the same
direction on disk-like planes.

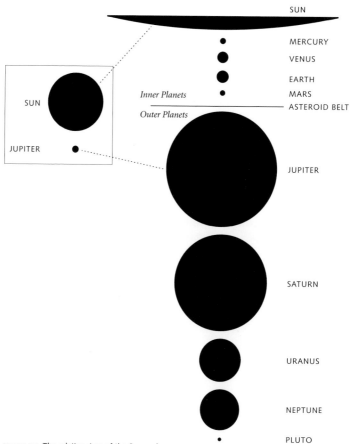

SUN

MERCURY

VENUS

EARTH

MARS

ASTEROID BELT

Inner Planets

Outer Planets

SUN

JUPITER

JUPITER

SATURN

URANUS

NEPTUNE

PLUTO

FIGURE 2.3 The relative sizes of the Sun and its planets.

chemical, and magnetic properties of the Moon's body and its heat. (The results will be explained in more detail in Chapter 3.)

Gravity measurements of the Moon and several planets were made by orbiting spacecraft, while seismometers on the Moon revealed its internal structure, including seismic boundaries and velocities. Figure 2.4 shows the structure of the terrestrial planets with their heavy metallic cores (not found on the Moon), their enormous silicate mantles, and their thin crusts consisting of very light elements like silicon, oxygen, and aluminum. This differentiation must have taken place rather early in the history of the Solar System, when the planets were still hot and went through a process that differentiated their density as in a blast furnace: the heavy material (e. g., iron) sank to the center, while the lighter material (e. g., slag, silicon, aluminum) stayed in the outer layers. Today, we can still observe signs of early magma oceans on Moon and Mars. We also see many early impact craters on both bodies that would not have survived if the differentiation had occurred later.

During the first 700 million years of the Solar System's existence, especially between 4.6 and 3.9 billion years ago, heavy meteorite bombardment occured. It began as chunks of material, big and small, escaped the gravity fields of the Sun and its planets. The resulting *accretional tail* orbited the Sun, the Earth, and the other planets, mostly in elliptical paths, and finally collided with the surfaces of these bodies. This period is called the "formative phase."

Interestingly, the Asteroid Belt between Mars and Jupiter never formed a planet of its own, probably because of the disturbing influence of the nearby giant Jupiter. Whenever orbiting asteroids come close to Jupiter, its immense gravitational force attracts them and disturbs their orbits, tearing them outside. The individual bodies, or planetesimals, in the Asteroid Belt consist of the

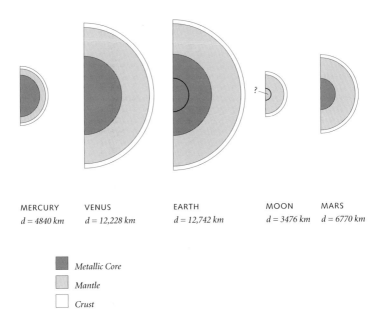

MERCURY   VENUS   EARTH   MOON   MARS
d = 4840 km   *d = 12,228 km*   *d = 12,742 km*   *d = 3476 km*   *d = 6770 km*

■ *Metallic Core*
▨ *Mantle*
□ *Crust*

FIGURE 2.4 The structure of the inner, or terrestrial, planets and our Moon. Seismologic studies revealed that the Earth's core is split into a fluid outer and a solid inner core. The chemical composition of the Moon's core remains unclear.

same material as the terrestrial planets (mainly silicates and iron). The earliest studies of iron meteorites at the beginning of the nineteenth century already led to the speculation that there must be a large quantity of iron in the interior of the Earth. Following Kepler's Laws and first terrestrial gravity measurements, the average density of the Earth was calculated to be higher than 5 grams per centimeter, while all known rocks have densities of only about 3 grams per centimeter.

Looking back at 7 to 8 billion years of pre-solar and pre-terrestrial history, it was the very beginning—the solar nebula, created by a supernova—which provided the iron for the Earth's core. The iron in our blood and the calcium in our bones and teeth also go back to this giant explosion.

chapter 3

The Formation of Earth and Moon

Because the Moon seems so close, circling the Earth at a distance of only about 400,000 kilometers, many scientists believed that the Moon had somehow "jumped out" of the Earth, possibly when the Earth was rotating faster. Such an "escape" seems physically impossible, but comparisons between rocks from Earth and Moon have revealed surprising similarities. This chapter will chart the development of the Moon and describe how scientists conduct Moon-based research to better understand the structure of the Earth.

Similarities and Differences

The Moon's chemical elements, minerals, and rocks are nearly identical to Earth's. Of course, there are important differences between the two bodies, especially regarding their outer surfaces. First, the Moon has no water and no atmosphere. It is too small and its gravity is too low for it to capture and keep any fugitive constituents, called *volatiles*. Second, the Moon has much less iron than the Earth, and third, its average density, only 3.34 grams per cubic centimeter, is even smaller than that of some meteorites.

All of these observations are explained by the fact that at the onset of the planets' final formative phase (around 4.6 to 4.5 billion years ago), many small and large orbiting bodies, so-called *planetesimals*, were on chaotic orbits as they attracted and collided with each other. When the Earth was nearly accreted, with its iron core already formed, it was hit by a huge body the size of Mars. The tremendous impact released large quantities of material from the mantle and tossed it into orbit around the Earth. The Moon formed out of these ejecta. At first, it revolved around the Earth on a much closer orbit than today, causing extremely forceful tides on our planet. The fast-shifting, powerful tidal stresses caused cracks in the Earth's crust to open or close, speeding up or slowing down any moving material. Today, similar processes can still be observed on the Moon. There, the tidal stresses caused by the large Earth are still powerful enough to generate moonquakes, preferably at perigee, i. e., when the Moon is nearest to Earth on its elliptical orbit.

Meteorites and other impacting bodies played an important role in the structural development of both the Earth and the Moon. A lateral impact during the accretion of a planet could either accelerate its rotation, as with Earth and Mars, or delay it, as was the case with both Venus and Mercury. Remnants of meteoritic rocks are found on the Earth and on the Moon. The first lunar rocks were formed 4.4 billion years ago out of a magma ocean, whereas the bigger, slower-cooling Earth produced solid rocks much later (about 4 billion years ago). Only some of these rocks survived, though many were destroyed or melted by dense lava eruptions and meteorite bombardments that took place on the Earth and the Moon.

The Moon, much smaller than the Earth, cooled early and eventually froze, forming solid rocks on the surface from a large magma ocean. This process occurred much earlier than it could have on Earth. . On the Moon, the oldest of these rocks are the *terrae*, which cover the far side of the Moon as well as the southwest part of the visible sphere. These rocks consist of a feldspar-laden (anorthositic) gabbro, a course-grained igneous rock. These light rock units endured intense meteorite bombardments, evidenced by a surface saturated with impact craters. On the warmer, more dynamic Earth, the early impact structures generated during the formative phase could not survive. The Earth was much warmer, tectonically active, and its surface was modified by both water and wind erosion.

While the Earth, with its large metallic core, developed a magnetic field rather early in its history, it is unclear whether the Moon ever developed a magnetic field of its own. Although some of the surface rocks found during the Apollo program are magnetized, the reason for this magnetization is still a mystery. There are, however, a number of theories. The magnetization could have been caused by a temporary internal field from a small, fluid iron core, or it might have been transferred from nearby Earth. The strong meteorite bombardment, not hampered by any atmosphere, could have produced shock magnetization.

Exploring the Moon

Figure 3.1 shows the front side of the Moon with the light terra landscape in the southwest and the dark and mostly circular dark (or *mare*) areas in the north and northeast. The *maria* (plural of mare) do not show the high density of impact craters of the terrae, hence they must be younger. Some younger impact craters, like

Copernicus and Tycho on the front side of the Moon, have ejected light material in the form of long streaks over the whole visible landscape. These ejecta consist of material thrown out of the impact crater and are distributed radially from the crater rim. Most of the material is deposited close to he rim, but impacts of large objects produce clots of ejecta, which become impact projectiles themselves and produce smaller impact craters. During the Apollo program, more than 2000 kilograms of rocks from terrae and maria were collected and investigated by radioactive dating and a number of other methods. The oldest terra rocks yielded an age of 4.4 to 4.2 billion years. The basaltic rocks from the mare area revealed ages between 3.9 and 3.2 billion years. The great circular mare basins were created rather late in the formative phase, around 4.2 to 3.9 billion years ago. Massive impacts, possibly by large, orbiting bodies, excavated huge basins several kilometers deep, which later—during a period from 3.9 to 3.2 billion years ago—filled with basaltic lava that rose from depths of 300 to 500 kilometers. The cooling lava formed dense rocks that still have enormous gravitational pull, especially in the huge, circular basins. Their density anomalies caused an asymmetry in the Moon's inertia, and the Moon developed a locked rotation. That is why it always shows the same face toward our planet.

Because of the long-time absence of any geological activity, we have to conclude that the Moon has been tectonically dead for about 3 billion years. The only known activity has come from a number of deep moonquakes induced by the Earth's tides. The strong, unsheltered radiation from the Sun during its day, about 14 Earth days, causes some thermal stresses and deformations on the surface. Observed by many orbiting spacecraft, the far side of the Moon is almost entirely composed of light terra material that

FIGURE 3.1 The Moon's body is locked in a
rotation that causes the front side to be
turned permanently toward the Earth. Note
the dark, basaltic *maria*, the light,
anorthositic *terrae*, and the numerous
impact craters with their long streaks of
ejecta. *(NOAO/AURA/NSF.)*

bears the scars of many early-impact craters. In fact, the deepest basin in the Solar System has very little lava inside and exists on the Moon's south pole. It is called the Aitken Basin and has a diameter of 2250 kilometers and a depth of 8.2 kilometers. No sunlight reaches the interior, and some scientists have postulated the presence of frozen water. Spectral studies conducted by the space orbiter *Clementine* in 1994 supported this assumption. In general, the differences between the front side and the far side in elevation, chemistry, magma (molten rock), gravity, and crust thickness are so strong that a lateral displacement of huge masses, possibly the onset of early plate tectonics, cannot be ruled out. If such a phase of global tectonics ever existed, it must have been very short because the small Moon cooled so quickly and became solid throughout. It seems that plate tectonics today is restricted to our planet alone, since Mars and Mercury are too cool today, and Venus is too hot.

The analysis of lunar meteorite impacts—which greatly decreased but continued even after 3.9 billion years—has been used to estimate the danger that meteorites and other impacting bodies present to our own planet. In the Earth–Moon system, the meteorite flux has been nearly constant for several billion years (statistically, large impacts occur every 100 to 200 million years). Toward the end of this century, space technology will hopefully be able to detect and destroy or divert any meteorites on a collision course with our planet.

chapter 4

The Interior of the Earth and the Role of Seismology

Before we embark on a thorough discussion of the Earth's chronological development in Chapter 8, it is helpful to have a closer look into the interior and structure of our planet. In this and the following three chapters, we will do just that, exploring the structure and properties of the Earth. Since seismology, the "queen" of geophysics, has revealed more information than all of the other geophysical methods combined, it will be introduced first.

Nearly all the boundaries and layers deep inside the Earth have been detected by seismology in the twentieth century. By observing seismic waves with all of their refractions, reflections, conversion, and scattering, we have been able to learn a great deal about our planet's structure.

John Winthrop, an eighteenth-century American scientist who studied earthquakes, is often credited with creating the discipline of seismology—the science of earthquakes. Today, seismology encompasses everything from an earthquake's energy and its focal mechanism to the stress which it generates, to the radiation of seismic waves, and on through their arrival at a seismic station. Special receivers, namely seismometers or seismographs, are used to measure these waves and monitor the displacement in three different directions and at various frequencies. Modern seismological observatories use a number of different instruments, often an array of instruments similar to arrays of radio antennas, to detect the magnitude and direction of arriving waves.

Seismic Waves

While seismic waves may be of devastating force near their sources, like all waves, they lose energy as they travel through the Earth. Figure 4.1 illustrates the principle of one type of seismic waves: the compressional, or P-waves (from Latin *prima unda*, "first wave"). Other waves—called shear or S-waves (from Latin *secunda unda*, "second wave")—travel much slower than P-waves. The shallower the earthquake, the more dangerous it is. Being body waves, P-and S-waves lose their energy to all sides. The fast P-waves oscillate in the direction of rays, an oscillation similar to those of sound waves; the slower S-waves have particles that oscillate perpendicular to the ray direction, somewhat similar to the propagation of light.

In addition to these body waves, there are two kinds of surface waves with rather complex motions, named *Love* and *Raleigh waves* after their discoverers. Love waves are similar to S-waves, but they run along the surface and only oscillate horizontally. Raleigh waves are more complex; they have a "retrograde" motion and consist of P- and S-contribution. All surface waves belong to the so-called *boundary waves*, a special wave type that is different from a body wave. Boundary waves are confined to a boundary

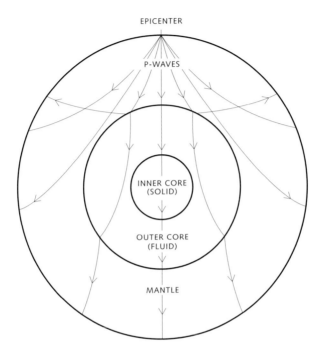

FIGURE 4.1 Seismic P-waves from an earthquake's epicenter passing through the Earth.

layer, such as the Earth's surface, and their amplitude exponentially decreases with greater depth. Generally, surface waves carry most of the energy and cause most of the damage on the Earth's surface.

Analyzing Seismic Waves

Earthquake observatories date back to the last decades of the nineteenth century. The first observatories housed huge seismographs, whose oscillators were equipped with styli that drew traces on soot-covered paper. In 1889, Ernst von Rebeur-Paschwitz recorded an earthquake in China by means of a self-constructed seismograph in Potsdam, Germany. Following this initial effort, instrumental techniques greatly improved throughout the twentieth century. After World War II, large seismological networks of broadband instruments were installed, mainly in the economically developed countries. During the Cold War, these networks recorded the sites of nuclear explosions. Eventually, the advent of computer technology allowed scientists to determine the location, magnitude, time, and mechanics of all medium and large earthquakes worldwide (and, after the 1950s, nuclear explosions as well). The largest increase in knowledge came from the concomitant development of detailed knowledge of the structure of the Earth and its various seismic velocities and densities.

Seismic waves are basically elastic waves that behave like light waves when they hit an appropriate boundary. According to Snell's Law, formulated by Willebrord Snell in the early seventeenth century, all waves change direction when they hit a boundary. In seismology, it is even more complicated: when a body wave hits a seismic boundary (where seismic velocity changes abruptly) it is not only reflected and refracted but, in addition, partially converted into another wave type—P into S, or S into P. A gradual increase in seismic velocities leads to a certain curvature of rays. S-waves do not enter the fluid core because the elastic modulus of a fluid (i. e., its shear modulus) equals 0. However, S-waves need the (elastic) shear modulus for their very existence.

To obtain an overview for relating seismic waves to the Earth's interior, seismologists use travel-time diagrams. Figure 4.2 shows a simplified version of such a diagram with only three individual seismograms (modern diagrams use more than a hundred individual, computer-derived, refracted, reflected, and diffracted travel time branches). From travel-time diagrams, seismic velocities and densities in the Earth are derived, as shown in Figure 4.3. In addition, details of the Earth's mantle, such as the presence of subsiding oceanic plates, are studied today by observing overlapping seismic rays from multiple earthquakes that are recorded at different seismograph stations. The latter method, called *seismic tomography*, is a recent development, having only become possible through the application of advanced computer technology. This technique is similar to tomography in the medical sciences, where a part of the body is "illuminated" for study. In addition, a multitude of seismic rays from more than 100 earthquakes and even more seismometer stations have been used to "illuminate" certain parts of the Earth's body. From so-called *seismic anomalies* in velocity, it is rather easy to derive *thermal anomalies*. They even show us the convection patterns in the Earth's mantle, the cold sinking lithospheric slabs, or the hot rising plumes from the core-mantle boundary at 3000 kilometers depth (see Chapter 13).

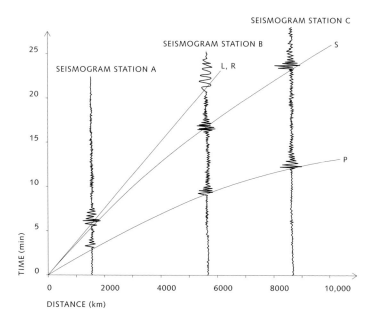

FIGURE 4.2 This travel-time diagram illustrates the increasing velocity of P- and S-waves (body waves), and Love (L) and Raleigh (R) waves (surface waves).

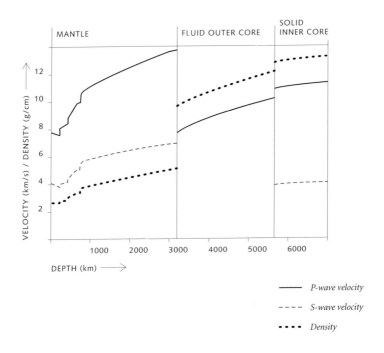

FIGURE 4.3 Seismic velocities and densities in the Earth's interior. The general increase of velocity and density is an effect of increasing pressure. Note the sharp drop in P-waves and the absence of S-waves in the outer core. Also note, in the shallow mantle, the jumps in P- and S-waves and density.

Free Oscillations of the Earth

Before we discuss the interior of the Earth in more detail, another exciting aspect of surface waves should be mentioned. Because their wave energy is limited to the surface, they do not lose much of their energy. It's the surface waves that are the great killers and can cause a tremendous damage around the epicenter of an earthquake. Since surface waves run in all directions, they cause a phenomenon called *free oscillations of the Earth*, where the Earth moves much like a big bell. This occurrence can generate several different modes of oscillations after a big earthquake (Figure 4.4 shows some examples), which are recorded by specially designed long-period seismometers, or gravimeters. The period of oscillation is independent from the location of the earthquake. With a period of nearly 54 minutes, the strongest mode is the "football mode" (where the Earth's oscillates elliptically, like a football).

While the amplitude of these oscillations is only a fraction of a millimeter, they reveal additional details about the Earth's structure. For example, scientists have learned that small deviations of the oscillations are caused by the Earth's rotation. Higher modes of free oscillation show slight differences between east–west and north–south oscillations. In addition, the density of the Earth contributes to the frequency of oscillations, especially in the low-frequency modes. The study of free oscillations is sometimes called *terrestrial spectroscopy* because analyzing the spectrum of the oscillations is somewhat similar to "real" (i. e., optical) spectroscopy.

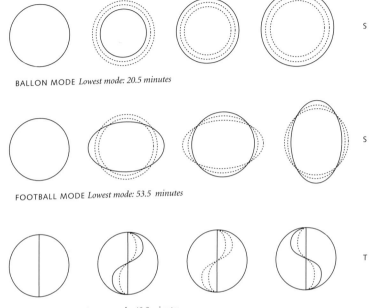

BALLON MODE *Lowest mode: 20.5 minutes*

FOOTBALL MODE *Lowest mode: 53.5 minutes*

TWISTING MODE *Lowest mode: 40.0 minutes*

FIGURE 4.4 Some simple forms of the Earth's free oscillations. The amplitudes are only a fraction of a millimeter. S = spherical, T = torsional oscillations.

The Earth's Interior

Figure 4.5 shows the spherical structure of the Earth with its metallic core in the center. The core is divided into a solid interior surrounded by a fluid outer core. The latter is home to our planet's magnetic field. Missing S-waves in the outer core suggest that this area is fluid, while the transformation from P- into S-waves on the boundary of the inner core indicate that the center is solid. The core is surrounded by a huge silicate-containing mantle, which, in turn, transforms into the thin crust, as displayed in an exaggerated way in Figure 4.5.

There are two boundaries within the mantle, and both are well described by seismological analyses. At a depth of approximately 410 kilometers, the so-called *phase transition* from olivine to spinel takes place. Here, the increasing seismic velocities cause iron, magnesium, and oxygen to group more densely.

The second seismic boundary within the mantle is located at a depth of about 660 kilometers, where spinel transforms to perovskite. This boundary shows a change in phase and material, and seismic velocities increase by about 8 percent. Perovskite is a dense mineral group made up of calcium, titanium, and oxygen. Although cold, fast-sinking slabs and hot, uprising plumes are able to cross this boundary (as explained in greater detail in Chapter 13), it is a barrier to convection. Both boundaries in the upper mantle cause a small jump in P- and S-waves as well as in density (see Figure 4.3).

There have been only small changes in these boundaries since the formation of the Earth. Changes in temperature during convection processes cause material to be transported laterally and vertically, which creates a potential for undulations and shifts

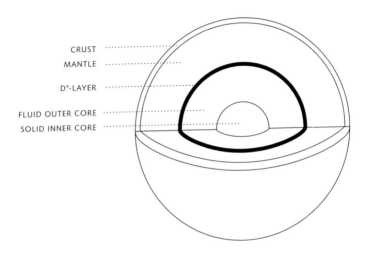

CRUST
MANTLE
D"-LAYER
FLUID OUTER CORE
SOLID INNER CORE

FIGURE 4.5 The structure of the Earth.

of the boundaries. Volcanism and hot springs attest to the fact that, today, the Earth is still warm and active. But it was much hotter and more dynamic in its early days, when heat was generated by various processes including the Earth's accretion, the formation of its iron core, the power of radioactive elements (especially uranium, thorium, and potassium), and, possibly, friction due to the tides caused by the nearby Moon. However, the Earth cooled rather fast in the beginning. The early fluid convection in the mantle transformed into a "solid-state" convection (a kind of creeping movement), and today, the Earth loses its heat mainly by conduction. The conduction of heat is a physical process without transport of material, and generally heat transfer in cool matter is much slower than in a hotter convection process. On Earth, we still have all three variations of heat transfer: fluid- and solid-state convection in the hotter parts of the Earth, and heat conduction mainly in the outer, cooler parts of the lithosphere. Unlike on Earth, convection is now dormant on the Moon and on Mars.

As will be shown in Chapter 10, plate tectonics also contribute to the continuing heat loss. Nevertheless, the current temperature at the center of the Earth's core is still around 5000 degrees Celsius, while the core-mantle boundary registers 3000 to 4000 degrees Celsius, with a significant temperature gradient between mantle and core. During the evolution of our planet, its solid inner core may have increased in diameter because of the cooling process, which transformed some fluid into solid material. This solidification possibly provides the energy for the dynamo action in the fluid outer core that is responsible for the Earth's magnetic field (as shown in Chapter 6).

Being very complex, with a variable thickness of several hundred kilometers, the transition from the fluid outer core to the lower mantle is one of the two most important boundaries on Earth. It is a chemical boundary, going from iron to silicate structures, as well as a physical boundary with very strong differences in temperature and density (see, again, Figure 4.3). Cold units from oceanic plates may subside and even reach the core-mantle boundary, while warmer units, possibly heated at the boundary, rise upward in the form of plumes. Seismologists call this boundary the *D"-layer*. Deciphering its structure and properties is a specific study area and one of the most exciting goals of seismic tomography.

The young science of seismology developed very fast in the twentieth century, successfully revealing the nature of the Earth's interior. However, seismology is not limited to the deep interior. The resources and variable structure of the Earth's crust, too, are an important target of seismic methods, as will be demonstrated in Chapter 11.

chapter 5

Rotation and Shape, Gravity and Tides

Centrifugal forces caused by the Earth's continual 24-hour rotation and its gravity determine the shape, or figure, of our planet. Both forces combined create the so-called *geoid*, determined by the hydrothermal equilibrium of the water subjected to gravitation and rotation. It is an *equipotential surface*, sometimes also called the *figure* of the Earth. To get a physical idea of the geoid, one can picture a canal that cuts through a continent from one seashore to the other, its water level representing the geoidal surface. The geoid is nearly an ellipsoid, or slightly flattened sphere. Its deviation from a sphere causes a bulge at the equator and a low at the poles with a difference of about 20 kilometers, a figure that has changed over time. The faster the Earth rotated, the more elliptical was its shape. This was apparently the case during the formative phase and the early Archean, when the Earth rotated in just ten hours and the Moon triggered dramatic tides on our planet, causing its shape to be much more elliptical. Slowly, the rotation velocity and the tidal friction decreased, and the Moon receded. Tidal friction itself is responsible for the decrease of the

Earth's rotation. Today, tidal bulges in the oceans appear twice per day, and the Earth has to rotate under these "brake shoes." Physics tells us that the orbital angular momentum (the product of angular velocity and orbital radius) of the Moon increases, so that the rotational angular momentum lost by the Earth is gained by the Moon's orbit. Or, to put it more simply, when the Earth's rate of rotation decreases, the Moon's orbit increases.

The equilibrium between the centrifugal and the gravity force was detected by Isaac Newton in the 1680s. He proved the physical validity of Kepler's Laws, which explain the orbits of the planets around the Sun and the moons around the planets. Newton also developed his universal *Law of Gravity*, which famously described an apple falling from a tree to the ground. Newton's Law held that two masses are attracted to each other with a force inversely proportional to the square of the distances between the masses. These Newtonian connections were not detected earlier because we cannot observe the described attraction in our normal lives. The number of Newton's universal constant of gravity G, or γ, is extremely small and can only be determined by complex experiments and instruments, such as the torsion balance. This highly exact instrument was invented in 1798 by Henry Cavendish in Cambridge, England. More than 100 years later, it was modified for practical field measurement by Roland Baron von Eötvös in Hungary. While Cavendish determined the exact constant of gravity, von Eötvös' measurements could specify gradients and curvature of the Earth's gravity field. However, measurements in the field lasted hours, and the torsion balance was eventually replaced by various types of gravimeters in the 1960s.

Describing the Earth's Shape

In order to describe the shape of the Earth, geophysicists use two types of figures. The first is the previously mentioned geoid, derived from physical measurements. It can be imagined as the surface of the sea (without winds, currents, or waves), continued below the continents by a system of channels. It is a plane of constant gravitational potential and has a much smoother shape than the real, rough surface of the Earth. The geoid is slightly higher below mountains and lower above depressions. It is affected by density anomalies in the subsurface. The gravity itself can be calculated from the vertical distance between two lines of the same gravity potential. The smaller the distance, the stronger the gravity. Orbiting satellites can determine the altitude and the shape of the geoid, although this is applied effectively only to the sea surface.

The second figure that is frequently used as a reference to describe the shape of our planet is purely mathematical. It is called the *reference ellipsoid* and describes the Earth's rotation with an axis to the poles being about 20 kilometers shorter than to the equator. It is a mathematical fit to the shape of the Earth and represents the fact that the general figure of the Earth is affected by the gravity directed toward the Earth's center, plus the centrifugal force of the rotating Earth directed away from its axis of rotation.

Both figures, the geoid and the reference ellipsoid, are used for specific geodetic and geophysical tasks. The difference between them reaches up to 100 kilometers, because density anomalies in the subsurface strongly affect the geoid and cause its undulations.

Tides

The gravitational pulls of the Moon and the Sun cause tides on our planet. The well-known ocean tides occur periodically as the Earth rotates. The gravitational pull from the Moon has about double amplitude of that from the Sun because the Moon, though small, is so much closer to Earth than the large Sun. If both bodies pull in the same direction, as happens during a full or new moon, we have spring tide. In this case, the water rises because the gravitational pull of the Sun and the Moon combine. This is not significant in the open sea, only about 0.5 to 1 meter, but it can rise up to several meters along the coasts in bays and inlets.

It is less known that the Earth's "solid" body also develops tides, which have periods like the ocean tides. Although we do not see or feel them, they reach 20 to 60 centimeters depending on latitude and position with respect to the average surface, or the reference ellipsoid. Every day, we move up and down by this amount, but only gravity meters can record the movement.

Rotation

The Earth's daily rotation operates under the drag forces of tidal friction, which steadily slow down the Earth's rotation. The largest friction is generated in the shallow seas, mainly in the large Bering Sea. Whenever scientists try to measure this change, they run into some difficulty because, over time, the Earth's shape has varied significantly. This causes problems calculating the rotational speed of the Earth for different geologic periods. As previously mentioned, physics proves that the Moon retreats when the

Earth's rotation slows down. Presently, its distance to the Earth increases by about 4 centimeters per year. This change can actually be observed in reflections of laser beams, since the Apollo astronauts have installed a laser reflector on the Moon.

Several small undulations are superimposed on the rotation of the Earth, accelerating or slowing the rotation or changing the axis of rotation. Geodetic and astronomical techniques enable scientists to measure changes down to fractions of seconds in rotation or centimeters of deviation of the axis. Astronomically determined latitudes, monitored at different observatories, show that the Earth's rotational axis has a free period of oscillations with a periodicity of about 14 months. This peak in the rotational spectrum is rather broad and has been called the *Chandler Wobble.* It is connected with a deviation of the Earth's axis of up to 6 meters.

While undulations of the Earth's axis became more known in the mid-nineteenth century, Seth Carlo Chandler, Jr., appeared on the scene. He was a prosperous merchant from Cambridge, Massachusetts, with enough time for a scientific hobby. Chandler repeated and analyzed some latitude observations and announced in 1891 that he had detected both a broad 14-month wobble and a sharp annual wobble. Although the initial reaction of the scientific community was rather unfriendly, his observations were confirmed, and the broad wobble was later named after him.

Even today, the exact mechanism that drives the Chandler Wobble is unclear. Especially the width of the spectral peak, probably caused by some damping mechanism, remains a mystery. Movements in the Earth's fluid outer core and earthquakes have been suggested as explanations, but the correlations are poor. It seems that currents in the ocean and/or atmospheric disturbances may play a dominant role. This shift of masses may really influence the rotation. The sharp annual wobble, on the other hand, can be explained more easily: It is caused by the sprouting of vegetation, particularly in the branches of trees in spring, and the falling of leaves in the autumn in the northern hemisphere.

The Earth's Magnetic Field

As everyone should know, the direction of Earth's magnetic field runs north–south and can be observed with a simple compass. The earliest evidence of humans being cognizant of these magnetic properties comes from Asia, where the Chinese used a magnetic needle as early as A.D. 100 to determine the direction of the North Pole. This knowledge reached the Mediterranean via the Arabs, and around A.D. 1200, the first iron needles were used on ships in the Mediterranean Sea. The Italians constructed a *bussola*, a device that first used a card balanced on top of a needle and later magnetic platelets swimming in a fluid. The Vikings had a kind of compass which they called *Leidarstern*, a magnetized rock capable of being turned around. Around the thirteenth century, a small deviation between magnetic and geographic north, called *declination*, was discovered in southern and central Europe, although it had first been detected in China around A.D. 720. In the fifteenth century, compasses were installed on many western ships. Columbus also used a simple compass for navigation when he went on his famous voyages westward.

Modern compasses appeared in Britain in the seventeenth century, where one was used by William Gilbert, the royal physician to Queen Elizabeth. Gilbert, who had enough spare time to study the magnetic field, was known for remarking "The whole Earth is a magnet." About 40 years later, a compass was used in exploration for iron ore in Sweden after it had become known that deviations from the north–south direction were caused by tiny amounts of iron in the rocks, mainly in the form of the mineral magnetite (Fe_3O_4). Later, the Swedes developed more refined instruments for geomagnetic prospecting.

Establishing a Physical Concept

Around the mid-nineteenth century, the ingenious Carl Friedrich Gauss developed new mathematical methods and established the basic theory of the Earth's magnetic field. He discovered that nearly the whole magnetic field is generated in the Earth's interior, while only a small variable part comes from an outer field (the ionosphere).

At first sight, our planet's magnetic field may seem easy to explain because it looks as if it were generated by a small permanent magnet in the center of the Earth. However, this explanation is physically impossible because every material loses its magnetic properties above a certain temperature. For the silicate rocks of our planet's crust and mantle, this temperature is the *Curie Point,* which is at 400 to 600 degrees Celsius, depending on the type of rock. This temperature range is reached at a depth of 30 to 60 kilometers; therefore, it follows that a hypothetical permanent magnet could not exist at greater depth, where temperatures are much higher.

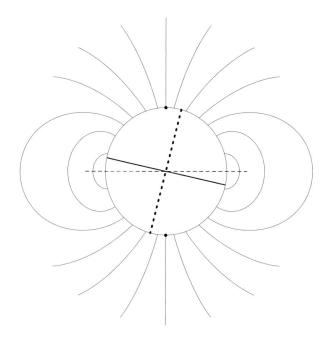

- ● MAGNETIC POLES
- ••••• ROTATIONAL AXIS
- – – – – GEOMAGNETIC EQUATOR
- ——— GEOGRAPHICAL EQUATOR

FIGURE 6.1 The Earth's magnetic field with field lines (simplified). The magnetic poles have a slight deviation from the geographic poles, or poles of rotation. The shape of the magnetic field looks as if a permanent magnet were near the center of the Earth; in fact, it is generated by a self-exciting *geodynamo* in the outer core. The magnetic equator is the equator relative to the magnetic poles.

The Geodynamo Theory

A magnetic field can be created in two different ways: by a permanent magnet (which, for the aforementioned physical reasons, cannot be the cause for our planet's magnetic field), or by a rotating electric current. It was Michael Faraday in the early nineteenth century who discovered the laws of *electromagnetic induction*, stating that a changing magnetic field of a moving or rotating magnet or a coil caused electric currents to flow, a phenomenon now called *Faraday's Law*. In the mid-nineteenth century, James Clerk Maxwell, the creator of the famous *Maxwell Equations*, discovered that every single flow of electric current is surrounded by a magnetic field. His equations establish a close connection between magnetic and electric phenomena.

We know that dynamos in power plants produce strong electric currents. We also know that cyclists transform mechanical energy into electric energy by pedaling. A dynamo attached to the tires creates electric currents for the bicycle's headlight beam. In these two examples, the magnetic fields around the electric currents do not play a dominant role. But somewhere inside the Earth we have to assume very strong rotating currents that generate a magnetic dipole field. The obvious place to look is the Earth's fluid iron core because it contains both the necessary conductors (iron) and the appropriate medium (fluid). Forty years ago, Edward Bullard and Walter Elsasser suggested that there really *is* a self-exciting dynamo, which they called *geodynamo*. The Earth's dipole field would thus be generated by a complex convective, probably helical, motion of fluid iron that is driven by heat generated through the slow upward movement at the core's solid–fluid boundary, or perhaps by a certain degree of radioactivity.

This dynamo theory is rather difficult to understand because it combines hydrodynamics and electromagnetics to "hydromagnetics," using complex systems of equations. But the mechanism is feasible, since the magnetic poles are located near the geographic poles and large-scale fluid motions in the outer core offer the only conclusive explanation for the magnetic field. Moreover, experiments with rotating disks and cylinders have demonstrated the feasibility of self-exiting dynamos.

Reversals of the Magnetic Field

The geodynamo theory explains many aspects of the Earth's magnetic field, such as its shape and strength, and the variation of declination and inclination—and yet it is not suitable to explain the magnetic reversal that has occurred regularly throughout our planet's long history. During such a reversal, the magnetic South Pole transforms into the North Pole, and vice versa. These periods are usually rather short; they mostly last only a few thousand years.

Over the last million years, the Earth's magnetic field has reversed six times, but paleomagnetic studies tell us that there were also long periods of about 20 million years without any reversal. The last reversal took place about 30,000 years ago. Our magnetic field is at least 3.5 billion years old, and for 98 percent of this time, it has been a stable dipole field. For about half of the time of its existence, it was a "normal" field, like at present, and for the rest of the time it was reverse.

Among the Solar System's inner planets only small Mercury, with its huge iron core, developed a strong magnetic field. Our

Moon may have had a small field at one time because of the magnetization of its old rocks. The huge outer planets all have strong magnetic fields; Jupiter has a normal field while Saturn's field is currently reverse. However, their fields are not generated by an iron core. Instead, the dipole fields are thought to be generated by fluid hydrogen, a real possibility considering the immense pressure found in these giants.

Paleomagnetism

The signature of the Earth's magnetic field has many applications for the various disciplines of applied geophysics. The magnetic field is a so-called *vector field*, i. e., it requires three parameters to be described, for example, *magnetic field strength* (H), *inclination* (I), and *declination* (D). The magnetization of rocks, caused by this field, is also described by three parameters, with the strength of magnetization (M) replacing field strength as a parameter.

Paleomagnetism, the scientific discipline used to determine magnetic parameters in rocks of different ages, requires complex experiments. Paleomagnetic studies are based on the fact that the magnetization of rocks that solidified from a molten stage is permanent, or *remanent*. Paleomagnetism plays an important role in plate tectonics research, since knowing the direction of the magnetic field at the time certain rocks were created can be related to the latitude of the sample at the time of its magnetization (see Figure 6.3). For example, a rock created somewhere in the southern latitudes drifted to the north, carrying its ancient magnetization with it. By determining the rock's age, preferably by radioactive dating methods (see Chapter 9), and assuming a dipole field, the drift velocity in the north–south direction can be

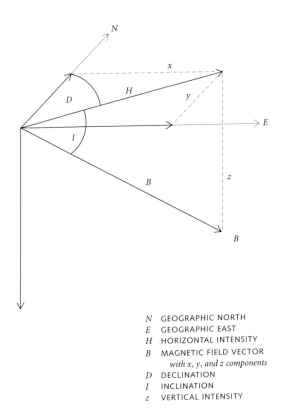

N	GEOGRAPHIC NORTH
E	GEOGRAPHIC EAST
H	HORIZONTAL INTENSITY
B	MAGNETIC FIELD VECTOR
	with x, y, and z components
D	DECLINATION
I	INCLINATION
z	VERTICAL INTENSITY

FIGURE 6.2 The parameters of a magnetic field.

calculated from the original *paleo-inclination.* Such a calculation, however, is not possible in east–west direction because of the north–south direction of the dipole field lines. But, at the very least, the *paleo-declination* tells us whether the continent has rotated.

Determining a Rock's Magnetization

Over the last half-century, the techniques for determining magnetization have improved significantly. Up until the 1970s, scientists could only specify the magnetization of igneous rocks, since these have strong magnetic fields. But eventually, methods for measuring the magnetization of sedimentary rocks were developed, creating a completely new branch of geology: *magnetic stratigraphy.* Layers of sedimentary rocks display the whole pattern of magnetic reversal, as correlations of borehole samples from all over the world revealed. Thus, the rate of sedimentation of many areas worldwide could be specified. Recently, scientists learned that not only magnetic minerals in the sediments cause their magnetization but certain iron-containing bacteria can have the same effect. These *magnetotactic bacteria* orient themselves in the direction of the acting magnetic field in a way that is similar to the orientation of normal magnetic grains. Many types of recent marine sediments display this biologically caused magnetization.

The most important of the newer techniques allowed the detection of the magnetization of the igneous ocean floors. This method, in turn, led to the concept of *seafloor spreading,* the corner-stone of *plate tectonics.* After World War II, many warships were transformed into research vessels. Equipped with powerful echo sounders, very long mountain ranges and ridges in the oceans were detected for the first time. Drum Matthews and his

*POSITION AT THE
TIME OF THE MAGNETIZATION*

*POSITION AT THE
TIME OF THE MAGNETIZATION*

CURRENT POSITION

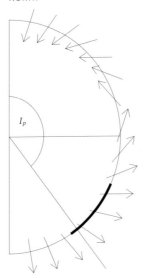

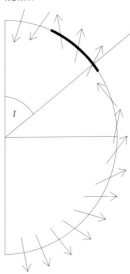

FIGURE 6.3 The Earth's magnetic field near the surface and the position of a continent. The paleo-inclination indicates a south–north migration of a continent. Arrows show the direction of the Earth's magnetic field. It is "frozen" in, and transported northward with, the continent, generally migrating with the drift velocity of plate tectonics (up to 15 centimeters per year).

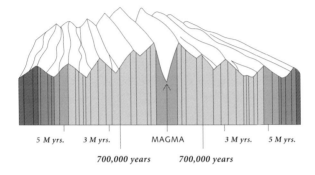

5 M yrs.   3 M yrs.   MAGMA   3 M yrs.   5 M yrs.

700,000 years   700,000 years

FIGURE 6.4 Magnetization of the Reykjanes Ridge ocean floor, south of Iceland. The observed magnetization parallel and symmetric to the ridge is obviously generated by alternating polarity of the seafloor. Normal magnetization: dark, reverse magnetization: light.

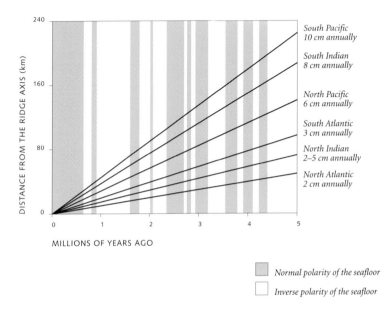

FIGURE 6.5. The polarity signature in various oceans displays the speed of seafloor spreading.

graduate student Fred Vine from Cambridge University had the idea to magnetically investigate the ridges. If volcanic, they should have a magnetic signature. With a magnetometer in tow, the researchers recorded symmetrically exiting signals on both sides of the ridges. There, the scientists were able to detect ridge-parallel stripes of normal and reverse magnetic polarization that matched those found in the boreholes taken on land. Because these samples had already been linked to a magnetic time scale, it was not too difficult to relate the magnetic stripes to certain epochs. Today, all ocean floors have been investigated, and the magnetic time scale has been greatly enlarged and analyzed back to Jurassic times (200 million years ago).

An explanation for the origin of the magnetic stripes was soon developed. Magma is continuously being expelled from the volcanic rifts. The molten material flows to both sides, cools down below the Curie point, solidifies, and assumes the polarity of the magnetic field at the time (normal or reverse). The seafloor thus spreads to both sides, carrying the remanent magnetization with it, creating a record of alternating polarization—the magnetic stripes. From the distance between magnetic stripes, or from their distance to the ridge axis, the spreading velocity of the seafloor can easily be calculated (see Figure 6.5). The growth rate ranges from nearly 10 centimeters per year for the South Pacific to only 2 centimeters per year for the North Atlantic. The seafloor spreading concept provides one important argument for the theory of the mobility of plates; another major concept will be described in Chapter 10.

chapter 7

Atom—
Mineral—
Rock

The atom (the word is derived from the Greek *atomos*, "not divisible"), was long regarded the smallest unit of matter. To understand the structure and evolution of minerals and rocks, however, we have to consider the components of atoms—neutrons, protons, and electrons. The type of a given atom, called *nuclide*, is specified by the number of *protons* (Z, also called the atomic number) and the number of *neutrons* (N) in the atomic nucleus. The sum of these is the *nucleon number* (A). The positive electric charge of the protons is balanced by orbiting negatively charged *electrons*. The number of protons in the nucleus, in turn, determines the chemical properties. At present, 264 stable, i. e., nondecaying nuclides are known. If a nucleus loses one or more of its electrons, it also loses its neutral property and becomes a *positive ion*; if it captures an electron, it becomes a *negative ion*. If two or more ions with different charges combine, a bond is created. There are different types of bonds, and the combination of bonds and ions determines the hardness of the next larger unit we'll consider here, the mineral.

Crystallization

In the geological sense, a *mineral* is a naturally forming compound with a characteristic crystal form. We call them *inorganic* to distinguish them from biological structures, but the difference is sometimes vague. For instance, calcium carbonate on the seafloor may have precipitated out of seawater, but seashells contain the same calcium carbonate. One may call it "biological" calcium carbonate, or—when it is created by an experiment—"synthetic" calcium carbonate. Earth's minerals are homogeneous substances, and *mineralogy* investigates the creation and development of minerals and their chemical and physical characteristics. Minerals mostly develop by a process called *crystallization*, in which a certain structure grows around very small seeds, or grains. This *crystallization structure* consists of a specific three-dimensional pattern of ions and atoms. Crystallization continues as long as there is enough space, and it is unhampered by other processes. As a result, we often find the larger crystals in caves, cracks, and faults. A familiar image of crystallization is the growth of ice crystals in water below 0 degrees Celsius. Rock-forming minerals also grow out of fluid materials, although their freezing (and melting) point is much higher, mostly between 600 and 1300 degrees Celsius, depending on their chemical composition.

Minerals in Crust and Mantle

Out of all naturally occurring minerals, only about ten form the known rocks in the Earth's crust and upper mantle. Most abundant are the compounds of silicon and oxygen, followed by several other oxides, i. e., compounds of oxygen and a metal. Carbonate compounds, mostly calcium or carbon with other elements, are also found in the crust. In total, crust elements containing oxygen constitute nearly 47 percent, those with silicon, 28 percent, and those with aluminum, 8 percent. The frequency of oxygen and silicon even increases from 75 percent to 93 percent if we add silica (SiO_2), e. g., in the form of quartz.

All minerals have a certain three-dimensional structure, which is based on the particular components of their atoms. For example, the basis of all silicates is a *tetrahedron*, a solid shape with four sides (see Figure 7.1). The small silicon ion is at the center, surrounded by four large oxygen ions. Such tetrahedrons combine with many types of similar structures to form rings, chains, streaks, etc.

In the upper mantle, the greenish mineral olivine dominates, making up more than 70 percent of the upper mantle's mineral content. It has an orthorhombic structure (i. e., a structure of three mutually perpendicular axes of different length) and—when oriented along streamlines—causes seismic waves to propagate with different speed in different directions in the shallow mantle. This phenomenon is also called *directional*, or *azimuthal*, *anisotropy*. It indicates that basically all anisotropic minerals are uniformly oriented, probably through creeping processes (as explained in Chapter 13). The olivine-dominated part of the upper mantle then transforms to a spinel zone at a depth of about 410 kilometers, and to the perovskite-dominated lower mantle at about 660 kilometers.

It seems very possible that the growth of clay minerals in the young Earth supported the first development of primitive life with the capacity for self-replication. As will be discussed in Chapter

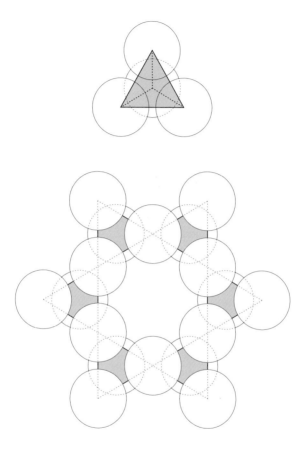

FIGURE 7.1 Two schematic views of the
structure of the silica tetrahedron (SiO_4).
The small silicon atom is surrounded by four
large oxygen atoms.

15, the slow growth of several crystals with complex structures and voids show patterns of replication. Clays have a very large surface area, voids and cracks in their structure, and make up about 50 percent of sedimentary rocks. They are the best candidates for initiating biological self-replication.

Rocks

Earth's rocks are mixtures of multiple units of minerals and make up the largest part of the Earth's body. The formation of both minerals and rocks occur more or less at the same time, and the appearance of a rock reflects the percentage and distribution of its minerals and the weathering involved. A rock's form and size depend on its evolution, and specifically on the depth (pressure), temperature, and time of its origin. The differences between the three basic types of rocks—igneous, sedimentary, and metamorphic—are not always sharp. Examples are displayed in Figure 7.2. Around 1790, the Scottish geologist James Hutton created a circle diagram that is, with some small modifications, still valid today. Figure 7.3 shows the complex paths and processes that constitute this rock cycle.

Igneous Rocks

Igneous rocks develop directly from the crystallization of silicate magma at high temperatures, usually between 600 to 1000 degrees Celsius, depending on the magma's composition. Mafic magmas, containing heavier elements like iron or magnesium, generally require 1000 to 1200 degrees Celsius to start melting. Silicate

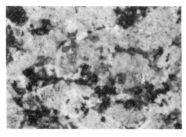

A. IGNEOUS ROCK

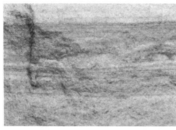

B. SEDIMENTARY ROCK

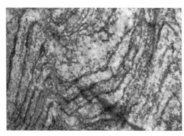

C. METAMORPHIC ROCK

FIGURE 7.2 The three basic types of rocks:
igneous, sedimentary, and metamorphic.
A: Solidification and crystallization from a melt.
B: Weathering, erosion, sedimentation.
C. Re-crystallization under high temperatures
and pressure.

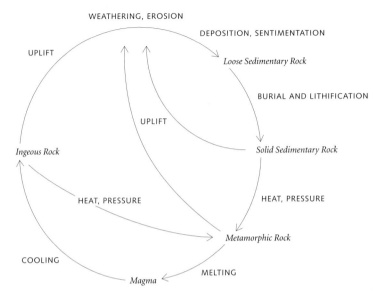

FIGURE 7.3 The rock cycle, after James
Hutton, around 1790.

magmas, containing lighter elements like silica, aluminum, and quartz, generally melt at 800 to 900 degrees Celsius, and if they contain water, the melting point can drop down to 600 degrees Celsius. There are generally two curves in a temperature-depth diagram that determine melting properties; the *solidus* is the upper limit of solid material. The *liquidus*, at lower temperatures than the *solidus*, determines the partial melting point in the presence of fluids. Solidified streams of lava and ashes both form igneous rocks. If cooling and crystallization take place at a certain depth and there is enough time available, crystals up to several millimeters long are created; these are sometimes found in granites and quartzites. Igneous rocks formed at depths are called *intrusive rocks*. They are found at all depths in the Earth's crust, possibly also in the mantle. If found at shallow depths, they are often accompanied by ores such as iron, tin, and copper. Intrusive rocks cool down more slowly than rocks at the surface, although they may be observed at the surface if transported upward by tectonic processes. Rocks from expelled ashes and flowing lava, on the other hand, cool very fast and form a fine-grained material called *effusive rocks*. Dark basaltic rocks, which form symmetric, mostly six-cornered columns when cooling and contracting, are effusive.

Sedimentary Rocks

Sediments consist of a wide range of *secondary rocks*, which are created mechanically or biologically from primary (i. e., igneous or metamorphic) rocks. Sedimentary rocks can also be chemically generated from seawater. We differentiate between *terrestrial* and *marine* sediments, also called *subaerial* and *subaquatic*. The terrestrial sediments come from mobile land masses at the surface—in avalanches or sliding slag heaps, say, or with the erosion of mountains or in high winds. Their mostly *clastic* (mechanically deposited) material changes from place to place, in contrast to the marine sediments that generally show a rather continuous deposition. Loose marine and terrestrial sediments can be transported over long distances.

Prior to becoming sedimentary rock, the loose material of sand, clay, gravel, or remnants of skeletons experience a solidification, or *diagenesis*, over time and due to the increased pressure from the growing weight of new depositions. The material is transformed into sandstone, slate, conglomerates, and carbonates, which, in turn, can be modified or destroyed by weathering and erosion. Some components experience a chemical modification. The most abundant *chemical*, or *organic*, *rock* is limestone, made up largely of fossil shells formed by biochemical precipitation of calcium carbonate that animals have extracted from seawater. The *evaporates* are composed mainly of gypsum, halite, and other salts crystallized from evaporated seawater in hot and dry climates.

The Earth's crust sometimes experiences *subsidence*, i. e., the slow sinking of a large part of the Earth's crust, generally caused by a cooling of intruded magmas or lateral extensional movements (see Chapter 12). With continuous subsidence, more than 10 kilometers of sediments can accumulate. For example, the Michigan Basin is nearly 30 kilometers deep. Here, the sedimentary layers are often interrupted by layers of igneous rocks because subsidence is often connected with extensional processes that favor volcanism. This is done by the reduced lithostatic pressure, bringing hot material over the pressure-dependent melting point curve (the *liquidus* or the *solidus*) into the melting range. The melt then rises and produces volcanism. Sediment basins often contain important resources, such as coal, lignite, oil, and gas.

At any depth, sediments can melt easily—even at 600 degrees Celsius if the material is sialic (rich in silica and alumina) and contains water. A subsequent tectonic compression may lead to a very complex structure with faults, uplift, melt, and inversion in the rock. For example, in the Rocky Mountains and the Alps, the formerly horizontal sedimentary strata have been folded, faulted, and sharply inclined by a subsequent tectonic compression.

Metamorphic Rocks

Metamorphic rocks develop out of igneous or sedimentary rocks under the influence of high temperature and/or pressure (depth). Through exposure to chemical or physical stress, igneous and sedimentary rocks may be deformed or re-crystallized, i. e., strongly modified ("metamorphosed"). They differ from their source rocks in composition because some elements either escaped or were captured in the hot or molten stage of a metamorphosis. Marble, which is metamorphosed limestone, or shale and slate, both of which are metamorphosed clay, are some typical examples. A large, deep metamorphosis is called *regional,* and this phenomenon is often associated with mountain formation. A more local metamorphosis, which can occur in the rocks surrounding a hot volcano, or *diapir,* is called a *contact metamorphosis.* The most abundant minerals in metamorphic rocks are, again, the silicates, a heritage from the high silicate content of igneous and sedimentary rocks.

Depending on the strength of the metamorphosis, a differently appearing or composed rock with *metamorphic facies* is created.

The so-called *blueschist facies* need high pressure and low temperatures, as they are created by the fast subduction of a cold oceanic plate. *Greenschist facies*, on the other hand, are formed at low pressure and high temperature, as was predominant at shallow depths in the young Earth. Average pressure and high temperature cause *granulite facies* that may change to *eclogite facies* at increasing pressure.

From the large variety of crystals, minerals, and rocks, only a few could be presented here. In addition to the three basic types of rocks—igneous, sedimentary and metamorphic—many tectonic, chemical, and physical processes change the structure and facies of a rock type. They all provide a record of geological events of the past and contain information about their origin and development.

Chapter 8

The Early Ages

The Archean

The formative phase of Earth and Moon took place 4.6 to 3.9 billion years ago. At the end of the formative phase, the Earth's first "real" geological epoch, the Archean, began. During this long period of time, which lasted up until 2.5 billion years ago, many geologically significant events took place. Many of the first rocks were formed during this period, although some already existed 4 billion years ago. These are observed in the oldest sections of the Earth, the relatively immobile cratons—substantial parts of the continental plates that have been almost entirely undisturbed since the Precambrian era. Yet only a few of these old rocks survived the strong meteorite flux and the intense volcanism of the early Earth. Some minerals (not rocks) can be traced back as far as 4.2 billion years.

On the Moon, the last giant impacts took place 3.9 to 3.8 billion years ago, when very large meteorites hit its front side, forming the large circular basins visible from Earth. At that point, 700 million years of strong flux came to an end.

The Earth and the Moon were undoubtedly hit by the same meteorite bombardments. (Our planet may have been plagued by a slightly heavier onslaught because of its stronger gravitational pull.) The larger Earth also cooled much more slowly than the Moon and, as a result, developed a strong internal convection and heavy volcanism. As our planet's surface was continuously reworked by all of these events, only a few rocks with greenschist facies survived.

At this time, the luminosity of the early Sun was about 25 percent less than today. But the Earth, though cooler, was not cold. It had an absorbing carbon dioxide atmosphere, similar to that of our neighbor planets Venus and Mars. Through volcanic eruptions, large quantities of carbon dioxide and water were exhaled, which led to the first rains and the earliest lakes and oceans. It is possible that some water reached the Earth by way of icy meteors, but it is difficult to estimate the relative significance.

The Emergence of Early Marine Life Forms

Shortly after the end of the strong meteorite bombing, about 3.8 billion years ago, first signs of primitive life began to appear. Today, we still find marine algae from that epoch incorporated in so-called *stromatolite limestones*, where they appear in the form of green-blue spherules, or spherical lamellae (see Figure 8.1). These first organisms developed a primitive photosynthesis, producing oxygen that combined with iron of valency 2 in ferric oxide (Fe_2O_3). This insoluble form of iron created the sedimentary *banded iron formations* (BIF). Today, these types of rocks, which

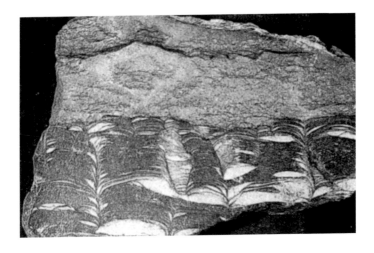

FIGURE 8.1 Stromatolite limestones from the Archean age show the first traces of life forms. The arc-like lamellae are believed to be remnants of colonies of blue algae.

were formed exclusively during the Archean and early Proterozoic ages, are the richest iron resources worldwide. Figure 8.2 shows an example of such BIF rock. It contains lamellae of chert (a siliceous rock) alternating with bands of ferric oxides (*hematite* and *limonite*) that exhibit a red and blue color.

Together with the production of oxygen, which was restricted to the oceans for at least 2 billion years, the amount of carbon dioxide, the main atmospheric gas, slowly decreased. This shift occurred because of the effects of rain: the carbon dioxide was removed, "washed out" from the atmosphere, and combined with calcium, hydrogen, and oxygen to form limestone, or coal and petroleum. At the seafloors, it mixed with mud and sediments to form carbonates.

Oxygen Enters the Atmosphere

In the early Proterozoic, about 2 billion years ago, oxygen entered the atmosphere. Up until that time, the oxygen, which was produced by photosynthesis in the sea, had combined with all the available iron until it reached the point at which it influenced the atmosphere and some continental rocks. Under the influence of oxygen (which then may have been only 1 percent of the atmospheric composition), *continental red beds* appeared, showing large amounts of ferric iron in the form of banded streaks (see Figure 8.3). Some oxygen on the continental shelves may have given rise to the first cell structures and, later, to oxygen-breathing organisms.

In the early Archean, the Earth was still hotter than today. Scientists estimate that, at least in the thinner lithosphere, temperatures were about 200 degrees Celsius higher. This caused a slightly different form of volcanism and produced the *greenstone-*

FIGURE 8.2 Red banded iron formations (BIF) from the Archean age today provide the largest iron resources worldwide. These deposits were formed as the photosynthesis of early organisms created oxygen, which, in turn, combined with iron particles to form solid ferric oxide (Fe_2O_3).

FIGURE 8.3 Continental red beds from the early Proterozoic age. Oxygen slowly entered the atmosphere and combined with iron to form ferric oxide (Fe_2O_3).

granulite belts and the higher-grade *granulite-gneiss terrains*, both of which are found in all old cratons. They are considered the first major contribution to the Earth's crust. The greenstone belts appear as elongated structures, but they have to be differentiated from the *orogenic belts*, which appear later, with the advent of plate tectonics. Greenstone belts are primarily composed of mafic-ultramafic lavas and some sediments, their green color being created by fine grained epitidote and chlorite. They stopped forming between 2 and 1 billion years ago, in the mid-Proterozoic age.

Even higher temperatures were needed for the origin of the so-called *komatiites*, magnesium-rich rocks from very hot ultramafic lava that often contains diamonds. Formed only in the Archean, they provided valuable ores for the kimberlites of today's South Africa. It seems strange that very near to the kimberlites there are other rocks that were created in a cold environment. Apparently, there must have been a temperature difference between about 1000 and 300 degrees Celsius at crustal depths of over 40 kilometers. This could be explained by a very narrow convection cycle, in which hot material was transported upward and cold material downward. An example for high crustal temperatures is early banded gneiss of granulite facies (metamorphosis at 600 to 700 degrees Celsius), while amphibolite facies (metamorphosis at 500 to 600 degrees Celsius) needs only slightly lower temperatures, showing a slight chemical similarity to the Moon's terra rocks.

The Proterozoic

The Proterozoic began 2.5 billion years ago. Life was established, but still on a microscopic scale. Multi-celled organisms began to develop only at the very end of this epoch, around 1 billion years ago. The continuous cooling of the Earth led to the growth of a rigid outer shell composed of the crust and the uppermost mantle, i. e., the *lithosphere*.

The Development of Plate Tectonics

By the time the lithosphere had developed, rifts and mobile belts, huge shelves, and large mountain chains began to form. About 1.9 billion years ago, the Skelefte mountain range developed in northern Europe. Its shallow subduction zone is still observable today by reflection seismology. Between 1.2 and 0.9 billion years ago, a very long mountain belt stretching from the Grenvillan area in eastern North America to the Svecofennian region in southern Sweden developed. Several continents accumulated to form the "super-continent" *Rodinia*. Thus, the motions of plate tectonics were well established. Two observations support this assumption. First, the appearance of *blueschist* in the middle of the Proterozoic is notable. Blueschist can only be created in the low-temperature, high-pressure environment of a steep oceanic subduction zone. Fast subductions carry cold material to great depth and pressure. Second, the formation of the first *ophiolites* can be observed. (Ophiolites are oceanic rocks thrown on land during a massive ocean–continent collision, and they are only found in strong convergence zones of mobile plates.) The super-continent Rodinia lasted at least from 1 billion to 650 million years ago. It then broke apart, but, nevertheless, it had lived longer than the later *Pangea* of Alfred Wegener's hypothesis. It was speculated that another super-continent in the Proterozoic preceded Rodinia, but scientific observations are as yet too vague. Another hypothesis assumes that there is a 500-million-year cycle, during which continents accumulate, stay together for some time, and disperse again. This theory is based on considerations of heat flow. If—possibly by chance—a super-continent has been formed, heat release from the interior is hampered by the thick continental lithosphere. Heat accumulates and forms magmas that rise, creating ridges or rifts, which finally lead to a spreading and dispersal of continents. The statistics might be poor, but the physical explanation makes sense.

The Formation of Today's Atmosphere

During the Proterozoic age, the atmosphere changed dramatically. Oxygen reached the present level. Carbon dioxide was, except for a very small fraction, expelled from the atmosphere and began to generate carbonates at the bottom of oceans and lakes. The noble gas argon increased as potassium decayed. However, the eventual dominance of oxygen-producing photosynthesis and the incorporation of carbon into the carbonates at the seafloor provided our planet with its present unique atmosphere, which is so much different from the carbon dioxide atmospheres of our neighbor planets Mars and Venus.

At the end of the Proterozoic, life forms had not yet evolved the hard skeletons that could have allowed us clearer fossil evidence of organisms in the epoch. However, that development came about very soon, at the beginning of the Phanerozoic age, approximately 570 million years ago. But before the "evolutionary burst" and the diversification of life could begin, the Earth experienced a terrible cooling, which transformed it into a kind of "snowball" (see Chapter 14).

Radioactive Dating

In Chapter 1, we briefly discussed the discovery of radioactivity and the work of Becquerel, the Curies, and Rutherford in the early twentieth century. Their research led to the discovery of many stable, and some unstable, elements as the century progressed.

The Chemistry of Unstable Elements

The decay of an unstable mother element into a stable daughter element is the basis for determination of absolute ages. New methods had to be invented to deal with all the newly discovered unstable elements, particularly the extraction of tiny elements from microscopic fluid, solid, and gaseous samples. These techniques were developed in several physics laboratories worldwide. Their acceptance by the geological community, however, proceeded slowly, though steadily, during the first 50 years of the twentieth century. Today, radioactive dating has developed into one of the most spectacular and promising disciplines of modern geology, relating the calculated absolute ages to certain rocks and tectonic events, and providing a stronger framework for describing tectonic evolution. As an example of present-day accuracy, the age of a 4-billion-year-old rock can be determined with three digits behind the decimal point! However, the needed techniques, such as mass spectrometry to separate elements and their isotopes, proved to be complex and difficult to develop.

Isotopes are different forms of one element, all having the same number of protons, but different numbers of neutrons in the nucleus. They began to play a special role in nuclear physics after scientists discovered that an unstable (radioactive) element disintegrates to form atoms with various isotopes of a different element. The analysis of these isotopes is carried out in different laboratories, specialized for greater and lesser ages.

During the second half of the twentieth century it became possible to analyze particles with a very small concentration of radioactive elements, such as the rare-earth element *neodymium* and its mother element *samarium*, which has an extremely long decay time (more than 100 billion years). This method developed into one of the most important dating techniques for large ages, together with the *uranium-lead* and *rubidium-strontium* calculations. Methods for determining small ages, i. e., for studies of climate or prehistory, require elements with short decay times. One procedure for dealing with such small ages is the carbon dating technique, which will be described at the end of this chapter.

Decay

The decay of radioactive elements is connected with the emission of particles and energy, which changes the number of protons (Z) and neutrons (N), and thus possibly also the nucleon number A ($A = Z + N$). From a mother element (M), a daughter element (D) is created. The daughter may be unstable, too, and then further decay until finally a stable daughter element is reached.

There are three kinds of decay processes, each of which is related to α-, β-, and γ-radiation. Through α-radiation, Z and N are lowered by 2 units, β- radiation increases Z and N by 1 unit, and γ- rays leave Z and N unchanged (see Figure 9.1). Mother elements decay exponentially with time, while daughter elements increase accordingly. The time it takes before only half of a mother element is left is called half-life (t_H). Based on the known half-life of different unstable elements, the ages of the oldest rocks and even the origin of the Earth can be determined.

Determining the Age

Radioactive age determination starts with the well-known *law of radioactive decay*, which states that the rate of decay dM/dt is proportional to the mother element M:

$$-dM/dt = \lambda M$$

where λ is the individual decay constant. The solution of this equation is $-\ln M = \lambda t + \ln M_0$. The integral form $M = M_0 + e^{\lambda t}$, shows the exponential form of the decay. The daughter element is $D = M_0 - M$, where M_0 is the initial concentration of the mother element.

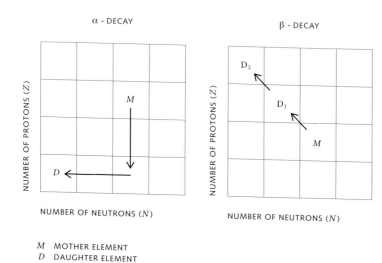

FIGURE 9.1 α- and β-decay.

Generally, we do not know the initial concentration M_0. We first express it in terms of the daughter D and then perform a calibration by introducing the non-radioactive daughter D_n, and eventually obtaining $D/D_n = D_0/D_n + (M/D_n) (e^{\lambda t} - 1)$.

D_0, the initial concentration of the radioactive daughter, is always rather small. The term $e^{\lambda t}$ can be approximated by λt. There are two possibilities for determining the time of origin t, as shown in Figure 9.2. We can plot D/D_n on the ordinate and either M/D_n or t on the abscissa. In the chart on the left, a straight line displays λt (and hence the time t) as a gradient; it also provides the *initial value* D_0/D_n as the intercept. On the right, we start with several samples and get several straight lines for different concentrations of the mother element. The most important experimental task is to determine D and M in a mass spectrometer (or by other techniques).

Age Determination in Closed Systems

Different minerals have different closing (blocking) temperatures, meaning that they solidify out of a melt at different temperatures. Once solidified, however, the mineral is a *closed system*, and no mothers and daughters can escape or be exchanged. The radioactive clock begins to tick and records the starting time t. Knowing the temperature T and the pressure of solidification p, we can even track the paths of a rock in a pressure-temperature-time diagram (Figure 9.3).

When radioactive dating techniques were first used, metamorphosis caused serious problems, and many age determination efforts established the time of the *last* metamorphosis where many minerals melted and mixed. Rocks are modified during metamor-

A

B

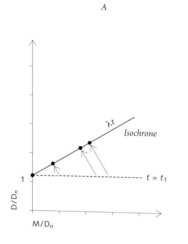

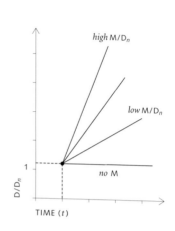

FIGURE 9.2 Two types of radioactive dating.
A: The gradient, or isochrone, provides the
absolute age *t*. B: The gradients show
various mother-daughter relations.

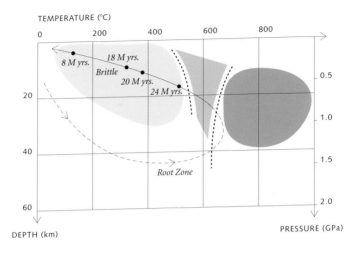

Low Grade

Medium Grade

High Grade

FIGURE 9.3 The circle-like path of a rock experiencing metamorphosis. The marked ages of solidification (24 to 8 million years ago) are confirmed by various minerals that formed closed systems at known temperatures.

phosis, and some minerals migrate, especially those with a low melting point. Using several rocks and concentrating on the minerals with the highest melting point, the metamorphosis and the origin of the rock can be determined. But if the rock was totally re-melted (in a general melt called *anatexis*), only the new solidification can be measured.

Over the years, scientists have learned that in order to identify the correct age of a rock or mineral, the sample has to be taken from a closed system where no migration of elements takes place. We can also apply the concept of a closed system to larger units, e. g., to a meteorite, a whole class of meteorites, a rock unit, parts of the crust, parts of the mantle, and even the whole Earth. The challenge is to collect representative samples. After observing that many samples from certain closed systems really do provide similar, reliable results and indicate comparable processes, scientists were able to develop even more advanced methods of age determination. For example, scientists have learned how to determine the formation times for various parts of the continental crust, because large parts were formed by a uniform and hot differentiation process of the mantle, providing similar temperatures and temperature histories. This created uniform and comparable ages of solidification. This process, in turn, led to a new method for trace element investigation and close study of enriched and depleted parts of the mantle. For example, the source of the *mid-ocean ridge basalts* (MORB) is depleted in incompatible elements, such as rubidium, barium, thorium, and uranium. The Hawaiian lavas and those from other plume sources (see Chapter 13) are less depleted on average, while the continental crust is strongly enriched in incompatible elements. Thus, any part of the mantle can be affected by crustal "contamination," for example caused by a delamination process (see, again, Chapter 13). Researchers have been able to discover the formation time of huge volcanic provinces as well as the origin and development of mountains or rifts. Even the collapse of the solar nebula and the formation of the Earth and other planetary bodies have been dated. Regarding the Earth's origin, it is interesting that the most different meteorites, such as chondrites from the outer Asteroid Belt (consisting of spherules of primitive matter) or achondrites from the interior Asteroid Belt (consisting of differentiated rocks and iron), all provide the same age of 4.6 billion years, and only an extremely small and special group suggests a slightly older age. The age of solidification is certainly the same as that of the accreting Earth and the adjacent planets.

Applications of Radioactive Dating Techniques

Radioactive dating methods are also put to use to assess ocean and continent drifts. While paleomagnetic studies specify the latitude of drifting continents (as discussed in Chapter 6), radioactive age determination reveals the time of their magnetization. From both observations (magnetization and time), the north–south drift of a continent can be clarified.

In addition, scientists apply radioactive dating methods to estimate the time at which a continent was split apart, or the time of a continent–continent collision. The approach of a continent or a *terrane* (see Chapter 13) can be further observed using paleontological methods to analyze the increasing similarity of fauna and flora before collision, which is just one more example of the steadily growing integration of the different geophysical and geological methods. This collaboration has led to our present knowledge of a mobile Earth.

Carbon Dating

Unstable elements with very short half-lives provide the basis for yet another spectacular form of radioactive dating methods. The carbon dating method has been developed to study materials that are no older than 100,000 years; the half-life of the carbon isotope ^{14}C is only around 5730 years. This method is the only one where the mother element ($^{14}C_0$) is available because it is continuously created in the atmosphere from the nitrogen isotope ^{14}N under the influence of radiation (even though some corrections have to be applied to the mother $^{14}C_0$, since it varies slightly depending on the strength of the magnetic field and the intensity of solar radiation). ^{14}C is absorbed constantly by plants through photosynthesis, and by humans and animals through respiration. Death, however, stops metabolism and respiration. The ^{14}C then begins to decay, starting the radioactive clock.

The age of many organic depositions like peat, moors, remnants of plants and trees, and skeletons of animals and humans can be specified using the carbon dating technique. Archeology, paleontology, and climatology have all contributed to and profited prodigiously from the exactness of this type of age determination. In climatology, for example, the relation of the stable isotopes of oxygen, i. e., the heavy ^{18}O to the lighter ^{16}O, provides reliable data on temperature because evaporation is temperature-dependent and prefers the light ^{16}O. International drilling programs in Antarctica and Greenland investigated air bubbles in ice cores and revealed a great deal about our planet's climate over the last 110,000 years, a period notable for its many advances and retreats of ice and glaciers. The very rugged appearance of the temperature curve of the last 100,000 years, as displayed in Figure 15.3, is one of the most exciting results of $^{16}O/^{18}O$ measurements in deep boreholes.

The human evolution during this period of time has been largely deciphered with the help of the carbon dating method. This was especially complicated because, for about 80,000 years, two species of hominids (the *Homo neanderthalensis* and the *Homo sapiens*) lived at the same time and sometimes even in the same regions. It was (and still is) a challenge for scientists to find out how and when the *Homo sapiens* arrived and the Neanderthals disappeared.

Paleontologists are perhaps the greatest winners of isotope geology. Fossils—their appearance, their mutation, and their extinction—can now be exactly dated, transforming the relative timescale, initiated in the nineteenth century, into an absolute scale. Moreover, an exact clock for the overall history of our planet has become available.

Plate Tectonics

At the beginning of this book, we mentioned that in the early twentieth century, a serious disagreement emerged between scientists with a "fixed" conception and those with a "mobile" view of our continents. At the time, both groups had good arguments to support their views. Alfred Wegener (1880–1930) collected the most compelling data among proponents of the mobile view. To make his case, Wegener pointed to the "fit" of the coastlines on the eastern and the western sides of the Atlantic, the continuity of mountain belts and many strata on both sides, and the correspondence of the records of ancient flora and fauna on either side of the ocean. Sir Harold Jeffreys (1891–1989), on the other hand, believed that the crust and the mantle of the Earth must be solid and rigid because shear (S-)waves were observed there. As explained in Chapter 4, these waves need a shear modulus for their propagation and, therefore, cannot exist in fluids such as water or the Earth's inner core. He believed that Wegener's concept of *sial* (a hypothetical layer of *si*licon and *al*uminum, which is much like the Earth's upper crust) swimming on fluid-like *sima* (a hypothetical layer of *si*licon and *ma*gnesium, which is very much like the Earth's lower crust) could not have been right.

From today's point of view, both luminaries were slightly wrong. Wegener greatly overestimated drift velocity and misinterpreted the depth of his transport mechanism. Of course, he could not have known at his time about the existence of a solid lithosphere on top of a viscous asthenosphere, a very thick and weak layer, which will be discussed later in the chapter. Jeffreys, on the other hand, could not have known that there is a viscoelastic zone, mainly our asthenosphere, in the upper mantle, and that it acts like an elastic body for high-frequency seismic waves but is still plastic enough for the slow, creeping movement of convection.

Twentieth-Century Research

All of this notwithstanding, Wegener, having postulated drifting continents after the collapse of the super-continent he named Pangea, is considered the great forerunner of plate tectonics. It took nearly 30 years, until the late 1960s, for the first clear indications of mobile plate tectonics to be recognized, but much of what Wegener hypothesized turned out to be true.

The first evidence came from the oceans. In the 1960s, Harry Hess of Princeton University suggested that seafloors separate along rifts in mid-ocean ridges, and that new seafloor forms by an upwelling of mantle material in these cracks—a hypothesis a few years later confirmed by Fred Vine and Drum Matthews of Cambridge University (as described in Chapter 6). Although a more general idea of plate tectonics had to wait for the advent of seismically detected subduction zones, where the seafloor disappears again, Harry Hess and the Canadian geophysicist Tuzo Wilson speculated, 30 years after Wegener, about the mobility of the lithosphere. However, no one could confirm the plastic or fluid-like layers on which the lithosphere was transported.

Thanks to the efforts of several geophysicists, there were many advances in convection theory and in our knowledge of the rheological (flowing and deformation) behavior of the Earth's interior that agreed with the seismological data obtained in the 1940s and 1950s. Seismic waves, both compressional and shear, were found to travel much slower at depths between roughly 100 and 300 kilometers. Calculation showed that temperatures nearly reached the melting point at these depths. This appeared to be a rather plastic and partially molten zone, later called the *asthenosphere*. The asthenosphere was not only deeper and less fluid than Wegener's sima, but it also allowed convection and movements over long periods.

The rigid zone comprising the Earth's crust and uppermost mantle on top of the asthenosphere was called the *lithosphere*. Depending on the temperature regime, the lithosphere reaches down to about 100 to 200 kilometers. Figure 10.1 shows the lithosphere and the asthenosphere for oceanic and continental environments.

Gathering Evidence

In Chapter 6, we mentioned that the magnetic stripes on both sides of the ocean ridges confirmed the seafloor spreading theory. The drift rate of the spreading ocean floor was then determined by

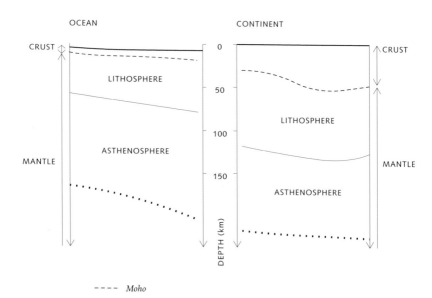

---- *Moho*

FIGURE 10.1. The rigid lithosphere, which includes both the crust and the uppermost mantle, lies on top of the creeping asthenosphere.

the distance of a known magnetic anomaly from the ridge axis, as shown in Figure 6.5. Later, deep-sea drilling determined the true age of the magnetic stripes, and it turned out that the radiometric ages corresponded with the magnetic ages.

Seafloor spreading thus provided one important argument supporting the concept of mobile plates. The new material thrown up by the seafloor spreading creates a new oceanic lithosphere, about 50 kilometers thick near the ridges and up to 150 kilometers at its edges. As it spreads, the depth of the lithosphere increases proportionally to the square root of time, following a deepening *isotherm* (line of constant temperature) of about 1250 degrees Celsius, which marks the transition to the asthenosphere. The asthenosphere, a thick, weak layer located at depths of 100 to 300 kilometers, had been discovered in the 1970s through temperature and melting point estimates and the measurement of a slight decrease in seismic velocities.

Slightly above the 1250-degrees-Celsius isotherm, at about 850 degrees Celsius, the ultramafic material of the oceanic lithosphere already gets "weaker." Below this depth, the "solid" lithosphere is not brittle enough to allow rupture; thus, it is the ultimate depth any earthquake-causing rupture can reach. This is similar to the characteristics of the continental crust, where the upper crust is brittle with earthquakes, while the lower crust is often ductile and creeping.

But what happens to the spreading oceanic lithosphere—where does it go? If the Earth did not expand steadily (at least one strange theory about this circulated in the 1970s), the ever-spreading lithosphere must somehow disappear into the mantle. Seismologists, hoping to find some earthquake traces in the lithosphere, attacked this problem with alacrity. In fact, earthquakes were found along an obliquely lowering zone, down to depths of nearly 700 kilometers near the South-American coast. This suggested that the oceanic plates were sliding down into the deep mantle. Study of the lineaments and the focal mechanism of earthquakes in the 1960s made clear that the whole oceanic plate *subducted,* not only off the South American coast, but also near Japan and in many other locations where moving oceanic or continental plates collide. New long-period seismometers and the signals from two very big earthquakes in Alaska and Peru in 1964 helped scientists like Jack Oliver and Maurice Ewing of the Lamont-Doherty Earth Observatory near New York to record a number of earthquakes along a downward "slab," and determining their focal plane mechanism (see Chapter 4). Oliver and his colleagues also coined the term *subduction* and postulated the first clear model of global plate tectonics. Calculation of temperature showed that the oceanic lithosphere carries its low temperature with it while subducting; it stays cooler than its surroundings and warms only slowly. The faster the subduction, the greater the depth of the cool and rigid plate. The faster the plate, the deeper it carries the low temperature, and the lower the temperature, the more solid and brittle is the material. The more brittle the slab, the more earthquakes will occur under stress. Also, the velocity of the moving oceanic plate near the ridges (calculated from the magnetic stripes) and near the coast (calculated from the observed rupture of shallow earthquakes) were in agreement when averaged over a couple of years, another confirmation of drifting. Additional proof of the drifting plates came from satellite geodesy and Very Long Baseline Interferometry (VLBI). These space-oriented geodetic techniques, using beams of lasers, light waves, or radar, can determine deviations between two points on

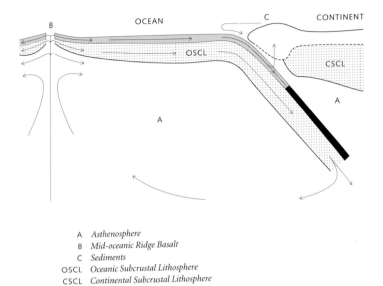

A Asthenosphere
B Mid-oceanic Ridge Basalt
C Sediments
OSCL Oceanic Subcrustal Lithosphere
CSCL Continental Subcrustal Lithosphere
----- Moho
Gabbroic Lower Crust
Eclogite Facies

FIGURE 10.2. Schematic profile of the lithosphere and asthenosphere of an oceanic plate, from its formation at a mid-oceanic ridge to its subduction under a continental plate.

the Earth's surface to 1 centimeter of accuracy. There is an international program to observe these changes of location worldwide, confirming conclusively the various drift velocities of the continental plates.

In the 1980s, seismologists discovered that the Earth's lithosphere is broken into a dozen or so rigid plates (Figure 10.3). They observed that the boundary of plates is marked by belts of earthquakes (Figure 10.4), and that there are three types of boundaries:

- *Boundaries of divergence* are found along the spreading ridges in the oceans. Earthquakes in these areas are mainly small and extensional.
- *Transform faults* occur in places where the plates slide past one another. These boundaries are found in the oceans (influenced by ridges) as well as inside continents (e. g., the famous San Andreas Fault in California or the North-Anatolian Fault in Turkey). Earthquakes here can be heavy and show a strike-slip movement, i. e., a movement horizontally and parallel to the strike of the fault.
- *Zones of convergence* are present where the leading edge of one plate overrides another. In ocean-continent zones of convergence, these areas mark the beginning of subduction and mountain building; in continent–continent collisions (like the Himalayas), huge compressive mountain belts are formed. Earthquakes along these zones can be extremely strong. Often subduction zones are steep, so earthquakes here are very deep, while they are only shallow at the other two types of boundaries (see Figure 10.4).

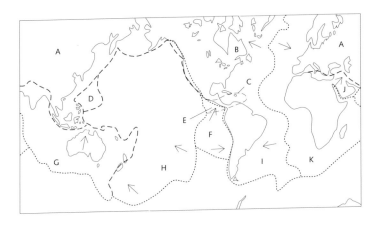

·········· MIDDLEOCEANIC RIDGE	A *Eurasian Plate*
	B *North American Plate*
– – – – SUBDUCTIONS	C *Caribbean Plate*
	D *Philippine Plate*
——→ DIRECTION OF PLATE MOVEMENT	E *Cocos Plate*
	F *Nazca Plate*
	G *Australian Plate*
	H *Pacific Plate*
	I *South American Plate*
	J *Arabian Plate*
	K *African Plate*

FIGURE 10.3 The lithospheric plates and plate boundaries. Subduction zones are zones of convergence with strong, deep earthquakes. Boundaries of divergence, where divergent spreading occurs (like at active oceanic ridges), show generally weak and shallow extensional earthquakes. At a "ridge jump" (a lateral offset of rift sections), a transform fault develops. These faults are mass-conserving boundaries with strike-slip displacements and weak, shallow earthquakes. Independent of ridge jumps, extended land transform faults, such as the San Andreas or the North Anatolian fault, may develop under the influence of oblique compressive stresses.

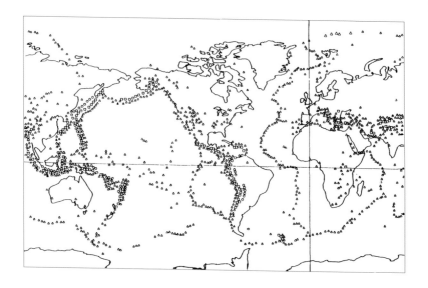

   ○ *Deep Earthquakes (> 300 km)*
   □ *Medium Earthquakes (70–300 km)*
   △ *Shallow Earthquakes (< 70 km)*

FIGURE 10.4 Earthquake belts indicate plate boundaries.

Drifting Plates

One important question remained: What drives the plates? Some energy certainly comes from the convection deep in the interior. In addition, oceanic plates obtain some of their energy from two other forces. First are the elevated oceanic ridges, where gravity, combined with lateral stresses, causes the famous *ridge push.* Second is the subducting lithosphere, its cool material (and some phase transition) producing a negative buoyancy, a so-called *slab pull.* Like a train, the oceanic plate is driven by two locomotives—one pulling and one pushing. Continents, on the other hand, are embedded in moving plates and are carried along only passively.

Many interesting phenomena can be studied at the rim of the Pacific Ocean, where the drift velocity is at maximum (over 10 centimeters per year), and where 90 percent of all heavy earthquakes take place. Strong compression occurs and small lateral components of stress cause very long faults parallel to the coastlines. Though most of the plate disappears in the subduction zone, a certain part of the oceanic crust that contains a lot of water melts at depths of 100 to 200 kilometers. The melt rises, and parts of the continental crust on the hanging wall are heated and thus also melt. The resulting magmas (called *andesitic,* from the Andes) are of a mixed type, mainly basaltic from the oceanic crust, but also showing a sialic material from the continent. These andesitic magmas cause a strong and often explosive volcanism, which can be observed nearly around the whole rim of the Pacific, giving it the name *Ring of Fire.* It should be mentioned that earthquakes do not *cause* volcanoes, but both phenomena are related to the subduction process.

How far do the subducting plates go down? Seismic tomography has recently solved that question. The slower plates penetrate the mantle down to about 660 kilometers, the boundary between the upper and the lower mantle. This boundary is characterized by a change of material, most probably from spinel to the denser perovskite. While slower plates seem to halt at this boundary and move horizontally, the faster ones cross it and even reach the D"-layer at the core-mantle boundary at a depth of 3000 kilometers. There, they may replace the material of rising plumes (which will be explained in Chapter 13).

Pangea and Beyond

By extrapolating from the present rift velocity back into time, on the basis of magnetic stripes and belts of ophiolites, the prehistoric positions of oceans and continents can be determined. Between 200 and 300 million years ago, all continents were assembled in one super-continent. This was Wegener's Pangea, a concept which was only much later supported by additional research. We know that Pangea existed as one big continent for nearly 100 million years, but then started to disintegrate. The disintegration began with a general extension and the opening of continental rifts, which were slowly transformed into narrow oceanic basins, similar to the process we observe today in the Red Sea. The Atlantic first opened in the south and then at the center. This opening seems to have occurred by pure chance, wherever an underlying branch of convection or weak section of lithosphere was present. The northern Atlantic separated later, about 65 million years ago, and the land bridge between the Danish Faroer

Islands and Greenland, is indicated by magnetic stripes and by the exchange of flora and fauna, was present some 20 million years longer. The Indian subcontinent "raced" north at a rate of nearly 16 centimeters per year and hit Asia around 60 million years ago. About the same time, the drift velocity of the big Pacific plate changed its direction by more than 20 degrees. Apparently, the dramatic collision that formed the Himalayas also extended its stress to the neighboring plates and initiated a re-orientation of the drift. The Hawaiian island chain shows this change of direction very clearly. We know that the volcanic Hawaiian chain is caused by a long-lived plume, which remains in nearly the same position while the Pacific plate moves over it with a velocity of about 10 centimeters per year in a northwest direction. Observing the sunken volcanoes to the northwest, a strong change of direction about 60 million years ago becomes apparent. Other island chains in the eastern Pacific show similar changes in direction, and we have to conclude that there was a change in drift around 60 million years ago.

Scientists have learned that drift and subduction really do change. This is especially true of intrusions, depending on the age of the plate and the tectonic stress field. Subduction could even retreat and later resume convergence with a new branch of the subducting lithosphere. In the eastern Pacific, subduction was not always perpendicular. There were times when the direction of the arriving oceanic plate was rather oblique, and other times when the subduction retreated. In general, however, the subduction off the coast of South America has been active for at least 300 million years.

In Chapter 8, we attempted to find out when plate tectonics began. It was suggested that the present type of plate tectonics with steep subduction zones must have started in the mid-Proterozoic, since at that time the first blueschist facies appeared. In the young Earth, subducting plates probably could not develop a negative buoyancy because they were too hot and possibly too thin and too short. Shallow, compression-dominated subduction, on the other hand, could have occurred rather early, as observed in very old compressive belts.

A certain modification of plate tectonics was suggested in the 1980s, based on the observation that today, and in the past, a number of *microplates* (or terranes) have split apart from a larger unit, drifting away with the moving seafloor, and eventually accreting to a new continent. These processes certainly contributed substantially to the formation of continents. We know that the whole of western and central Europe, Alaska, and parts of Siberia were formed by the compressive docking of terranes, a phenomenon that will be discussed in greater detail in Chapter 13.

chapter 11

The Crust
of the Earth

In the last chapter, we examined the two basic units of plate tectonics: the weak asthenosphere below the rather solid lithosphere. The solid lithosphere consists of the Earth's crust and the subcrustal lithosphere that belongs to the Earth's mantle. Both units are divided by the *Mohorovičić discontinuity,* or *Moho* for short. In this chapter, we will discuss the structure and composition of the crust and different tools that scientists use to study it.

The Moho

The Moho was first discovered in 1909 by Andrija Mohorovičić, a Hungarian-Croatian seismologist who observed travel times from an Alpine earthquake by means of a number of simple seismometers. These travel times (see Chapter 4) showed a strong change of seismic P-velocities at a depth of 20 to 30 kilometers. Mohorovičić recognized that this change in velocity might mark a change between the crust and the mantle. This, in turn, led to recognition of several important ideas. First, Mohorovičić's discovery suggested that the crust was very different from the mantle and that they were separated by a rather sharp boundary. Second, it suggested that the crust contained light "sialic" material with lower seismic velocities and lower densities than the underlying mantle. Third, it led to the recognition that the Moho—a boundary of mineralogical, elastic, and chemical properties that can be observed by seismic and gravity measurements—is one of the two most significant boundaries of the Earth (the other is the core–mantle boundary).

The Crust

The Earth's crust varies tremendously in thickness and in the ways it affects the Earth's surface. Though the solid part of the oceanic crust is only about 5 to 7 kilometers thick and rather homogenous, the continental crust shows a big variety of rock units and depths ranging from 20 to more than 70 kilometers. In young mountain belts like in the Tibet-Himalaya region or the high plateaus of the Andes, the continental crust reaches a maximum depth of more than 70 kilometers. The only known thicker crust is on the far side of the Moon; it reaches 120 kilometers.

The Earth's crust also varies in its composition. Figure 11.1 gives an overview of different types. On top, there are the sediments (see also Chapter 7), which may reach a maximum depth of more than 10 kilometers in basins and depressions.

Sediments are generally considered part of the crust, although some researchers differentiate between the *sedimentary* crust on

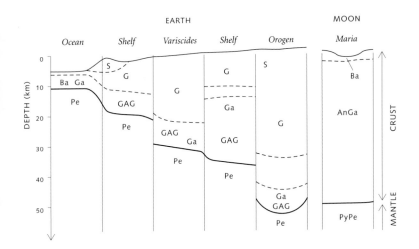

FIGURE 11.1 Various types of crust, revealed by P-velocities (v_P).

top and the *crystalline* crust below. The latter primarily consists of metamorphic rocks along with smaller quantities of igneous rock like granite and gabbro. The homogeneous oceanic crust mainly consists of basalt and gabbro, both igneous rock types that were generated near the volcanic ridges and transported to the sides. Sediments become more prevalent and generally thicker with the crust approaching a continent. The various rocks of the continental crust consist mainly of oxides of silicon and aluminum, i. e., 60 percent is SiO_2 and 15 percent Al_2O_3. As mentioned before, the high concentration of silicon and aluminum led Alfred Wegener to postulate a *Sial* layer on top of *Sima* (silicon and magnesium). Today, we consider silicon and aluminum a dominant part of the *upper* crust, while the *lower* crust is primarily of a more basic chemical composition, containing some heavier elements like iron and magnesium, and rocks with gabbro or granulite facies.

Formation and Development

The crust of the terrestrial planets, including Earth, was created out of the underlying mantle by a global differentiation process. Huge modifications were caused by magmatic flooding, extrusions, and various forms of volcanism. A certain amount of recycling takes place in Earth's subduction zones. The active ridges in the oceans still produce huge amounts of basaltic magma out of the underlying mantle, and today, these ridges are the dominant producers of new basaltic material. There are also relatively minor outflows of basaltic lava from volcanic areas like the island of Hawaii, Mount Etna, and huge regions of ancient volcanism in India and Siberia.

Andesitic lava, a type of mixed lava, occurs in mountains over subduction zones as well as in many other continental areas where deep tectonic processes, usually subduction and intrusion of oceanic units, take place. Famous locales of this often-explosive volcanism are the Andes, Japan, the Philippines, and mountains such as Saint Helen's, Vesuvius, and Stromboli. All of these volcanoes are near active plate boundaries. Whenever lava rises and comes into contact with either ocean or ground water, explosions will follow. The combination of continental material, e. g., silicon, triggers a chemical reaction that, while not well understood, is believed to make this "mixed lava" particularly dangerous. Mafic lavas from the mantle have a higher melting point than normal crustal material and will melt crustal rocks easily. More silicic magmas, such as granite plutons, domes, or intrusions, can also be produced in the mantle under special conditions, but they mostly come out of the lower crust. Under the influence of heating—from rising magmas, plumes (see Chapter 13), or just by decompression (Chapter 10)—crustal material may melt. In a so-called *anatexis*, melting plus fusion of various minerals and rock units take place, creating the silicic magmas. In general, volcanism, very strong in certain time periods, has contributed greatly to the steady growth of the Earth's crust.

Exploring the Crust

In Chapter 4, we introduced seismology as the most successful method for the general study of our planet's interior structures. Here, we will review the present applications of seismology as it is employed to explore the Earth's crust in particular.

The 1920s saw the creation of the *seismic refraction method*. Similar to Fermat's principle for light waves, seismic waves also tend to find the fastest way between the source (shot) and the receiver (geophone). The fastest way is usually not along a straight line, but along a ray that makes use of the increasing velocity with depth. If we have a low-velocity sediment on top of a higher-velocity consolidated rock unit (such as an older sediment, or a consolidated or crystalline layer), the seismic ray will follow the upper boundary of the lower layer. The wave is called the *head wave* and forms the basis for refraction seismology. From the travel times, one could observe the dip of the lower layer and any disturbances or changes in its velocity—a direct continuation of Mohorovičić's method. It was German seismologist Ludger Mintrop who first applied this method commercially, searching for salt domes and salt structures in Texas, and especially for their oil-bearing flanks. Refraction seismology is a relatively simple method. It does not need refined amplifiers, nor does it require summing up the output of a number of geophones, a procedure called *stacking*. Although refraction seismology was overtaken by *reflection seismology* in the 1940s, it is nevertheless still used for reconnaissance and crustal studies.

In the 1950s, refraction was greatly refined and became a tool of *explosion* or *controlled-sources seismology*. This active seismological method is used worldwide for crust studies. Instead of waiting for the next real earthquake, seismic energy is generated by explosions, vibrators, or airguns. Similar to the technique developed for commercial refraction seismology, seismometers are replaced by small short-period geophones that transform the arriving seismic signals into electric pulses. In general, the length of observation lines (the *profile* or *line*) is a few hundred kilometers, with geophones dug in every 3 or 5 kilometers. In the 1950s and 1960s, the seismic signals were generated by quarry blast explosions or by big underwater explosions in lakes, which provided powerful, cheap, and relatively safe sources of low-frequency signals. The seismic signals along the observation lines are first processed in the field and later, after analog or digital processing, are filtered, stacked, and interpreted in various laboratories. As in earthquake seismology, signals are plotted according to their distance from the source and their travel time to the geophone.

In the 1970s, large multinational projects were initiated in the younger Alps and the mountains of Scandinavia. Below the Alps, a crustal root about 55 kilometers deep was found, a result that was only roughly predicted by isostatic estimates and gravity observations. Below the Scandinavian Caledonides, no root was found, although gravity measurements had predicted its presence. Along with the improvement of the recording technique, which involved overlapping receivers and the stacking of signals, smaller shots from boreholes or even vibrators were applied, using several shot points along the line. The late Soviet Union went even further, detonating underground nuclear explosions to provide powerful signals along profiles more than 2000 kilometers in length. The resulting high-frequency arrivals penetrated large parts of the mantle and provided better resolution than the low-frequency signals from natural earthquakes. Although the resolution for the crust improved only slightly, several new boundaries were detected in the mantel, but their significance is still unknown. The upper mantle turned out to have some weak boundaries, such as the *Lehmann discontinuity* at a depth of 200 kilometers (discovered by Danish seismologist Inge Lehmann), where seismic anisotropy seems to terminate.

After such experiments are carried out, there are several mathematical methods for deriving a velocity-depth model with seismic boundaries from overlapping travel-time diagrams. Over time, scientists have learned that the strongest seismic boundary is almost always the Moho at the bottom of the crust because of its change of velocity, density, and chemistry. The most powerful signal is the wide-angle reflection from the Moho (called a P_MP event). The P_MP is a real P-wave reflection from the Moho. For near-vertical rays, its energy is rather low, and we can detect it only with complex receiving and processing techniques. However, for "wide-angle" distances, say, 60 to 90 kilometers from the source, the "critical angle" is reached, and nearly all the seismic energy is reflected. In optics, this is the angle of *total reflection*; in seismology, the physical conditions are slightly more complicated because there is a slight conversion from P- to S-waves. In any event, energy in the wide-angle area will always be very strong.

In addition to the reflection, we will receive a refracted wave from the Moho. For long observation distances, say, 120 to 200 kilometers, the so-called head wave from the Moho can be observed. For crustal depths of 30 to 40 kilometers, one can imagine that it takes some time (and distance) for a wave to travel through the crust down to the Moho, run along the Moho, and come up again. This wave is called the P_n-wave. Over very long distances, it "dives" slightly into the uppermost mantle because of a small velocity gradient that is the product of increasing lithostatic pressure.

Sometimes the Moho is not a sharp boundary, but rather a transitional zone. (This is frequently seen with deep Moho.) Shallow crusts, on the other hand, tend to have sharp (so-called first-order) boundaries. Often, seismic boundaries are detected inside the crust, which show mostly small changes of elastic parameters (see Figure 11.1). One of these intracrustal boundaries is the *Conrad discontinuity*. This boundary cannot be observed worldwide but seems to be present in young crusts and sometimes in shield areas like in Canada. It marks the boundary between sialic upper and mafic lower crust, where velocities increase from around 5.9 to 6.3 kilometers per second to more than 6.5 kilometers per second. However, the various types of crust significantly differ from each other. Young crusts are generally shallow and can be divided into two main layers, while many crusts in old shield areas have a greater thickness and an additional layer with velocities of more than 7 kilometers per second at the bottom. From the seismic velocity-depth models even the petrology of the layers can be estimated. This has been perhaps the biggest achievement of the seismic refraction method.

However, for all of its uses, the refraction method is not without limitations. For example, it cannot resolve faults, intrusions, low-velocity zones, and small-scale inhomogeneities. For detecting the fine structure of the crust, another seismic method, *near-vertical reflection seismology*, is applied. Seismic waves hitting a seismic boundary are reflected back to the surface just as acoustic waves from an echo sounder are reflected from the sea bottom. Because the difference in energy between seismic rays from a shallow interface and those from a very deep interface can be extremely large—reaching a factor of up to 100,000—reflection seismology was adopted more slowly by the scientific community. Only after the addition of a feedback mechanism for compressing the first strong arrivals was it possible to use this new method.

Since the 1930s, exploration companies have looked for hydrocarbons by searching for structural traps in the sediments. In the

1940s and 1950s, their technique was improved by the introduction of filtering, stacking, and common depth-point techniques (where a reflection point is attacked by seismic rays from various angles). Digital recordings and more sophisticated processing techniques followed in the 1960s and 1970s. Today, three-dimensional recordings of an area, using a multitude of shots, vibrators, airguns, and receivers for marine and terrestrial surveys, provide detailed pictures of sediments. Even the seismic velocity of sediments is monitored, and oil companies spend up to 95 percent of their exploration budgets for reflection seismology. Figures 11.2 and 11.3 show simplified pictures of on-shore and off-shore surveys.

Given the success of reflection seismology for sedimentary structures, it became obvious that it should also be tried for crustal studies. Though reflection methods are more complex and expensive than refraction studies, sporadic tests conducted in the 1950s and 1960s in Canada and Germany were promising. In 1975, eventually, the first big crustal reflection program, COCORP, was launched in the US, with similar groups established in other countries shortly thereafter. In Great Britain, the BIRPS consortium applied marine techniques, using airguns to provide the seismic energy that is received by hydrophones (see Figure 11.3). This offshore research is often preferred because the receivers are towed behind a ship and need not be carefully arranged on the ground. This important difference makes the experiments about 20 times faster and more economical than those based on land. Figure 11.4 shows an example from DEKORP, the national German research group, in the Variscan Mountains, where a laminated lower crust terminates with the Moho at a two-way reflection travel time of about 10 seconds. Usually, two-way travel time is displayed and plotted on the vertical

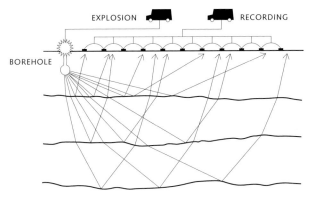

FIGURE 11.2 Reflection seismology on land. Seismic signals are generated by explosions and recorded by geophones.

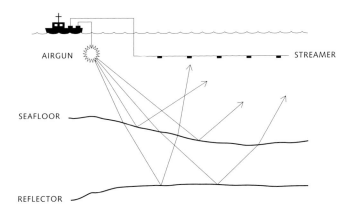

FIGURE 11.3 Reflection seismology at sea. Seismic signals are generated by airguns and recorded by pressure-sensitive hydrophones in a streamer, towed behind a ship.

axis because it is the direct and primary output of the reflection measurements. For an average seismic crustal velocity of 6 kilometers per second, a two-way travel time of 10 seconds corresponds to a depth of 30 kilometers ($z = V \times t/2$; where z = depth; V = velocity; t = two-way travel time). This lamination appears only on the bottom of rather young and shallow extensional crusts. Apparently, a recent extensional heating that occurred in connection with extensional stresses has ironed out any former irregularities, creating a horizontal layering of contrasting impedance (i. e., rock units with differences in seismic velocity and density).

While reflection and refraction programs are often completely separate, some reflection programs do incorporate refraction observations. Signals from explosions are recorded at great distances with mobile geophone stations so as to capture both reflections *and* refractions. This hybrid recording technique combines the advantages of both methods: Researchers get to study the fine structure through reflection, and can observe the velocities through refraction. Today, the Moho boundary is almost always observed by both methods. Furthermore, gravity surveys are often combined with seismic investigations or used in areas where no seismic profiles are available. The Moho can be determined because it shows the strongest difference in density within the outer part of the Earth and dominates the gravity signals. By modeling the gravitational signal from a crustal density model and adapting it to the observed gravity, undulations of the Moho can be specified.

DEKORP IIS
Saxothuringian-Moldanubian

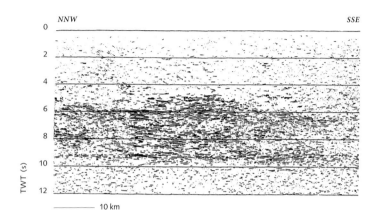

FIGURE 11.4 Example of a crustal record section from DEKORP in the Variscan Mountains, Germany. Note signs of faults in the otherwise transparent upper crust and massive lamination in the highly reflective lower crust. The lamination was presumably created by shear movements in an extensional stress system when the mountains collapsed. The vertical scale shows the two-way reflection travel time (TWT) in seconds. Intervals of 10 seconds correspond to a depth of roughly 30 kilometers.

The Depth of the Crust

It is known that the crustal depth varies (see Figure 11.1 again), depending on evolution and the latest tectonic events. In the old shield and platform areas of central North America, Asia, and Africa, the crust is between 40 and 50 kilometers deep. As mentioned before, it reaches even more than 70 kilometers in young compressed mountain belts, such as in Tibet or the Altiplano region of the Andes. The younger rim of North America shows a lesser crustal depth, reaching in the Appalachians to about 35 to 40 kilometers and in California to about 30 kilometers. The Scandinavian crust is rather thick, which is not surprising since it belongs to the old Baltica craton. The age of the Baltica craton is Archean in the very north, where it is more than 3 billion years old. It gets gradually younger to the south, first displaying rock structures from the older Proterozoic (around 2 billion years) and then from the younger Proterozoic (around 1 billion years). This succession of ages in one direction is a manifestation of continental growth, where younger continental plates and terranes continuously and discontinuously accrete. In a similar way, North America (which was joined with Eurasia until about 200 million years ago) collected continental plates and terranes around its oldest nucleus in present-day Canada. Figure 11.5 shows the crustal thickness in North America as obtained by a combination of reflection and refraction methods.

The Deformation and Flow of Matter

The continental crust's *rheology*, or deformation and flow of matter, is very complex. In Chapter 10, we mentioned that earthquakes are restricted to a certain shallow depth (except those in

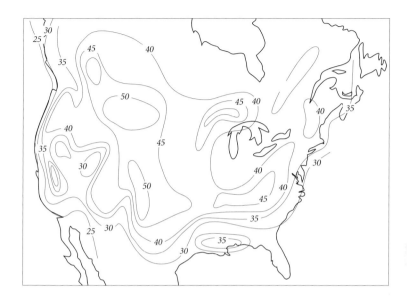

FIGURE 11.5 A map of crustal thickness (Moho map) for North America—contour interval: 5 kilometers. Note the larger crustal thickness (> 45 kilometers) in old shield areas and young mountain belts, and the smaller thickness (< 35 kilometers) in young and coastal areas.

subduction zones). For the continental crust, a depth of 20 kilometers seems to be the lower limit for generating rupture and earthquakes. If tectonic stresses become larger than the strength of the rock unit, which is between 0.1 and 0.4 megapascals, depending on depth and water content, the rock unit will rupture. Below a depth of 12 kilometers in young crusts and about 20 kilometers in old crusts, the material becomes weak, and young and warm crusts often show signs of creep, as seen in the sub-horizontal creep of young extensional crusts. Moreover, many creep-and-rupture experiments show that most rocks at temperatures of less than 300 degrees Celsius remain solid and brittle. They transform into a more ductile, or plastic, state at higher temperatures, as they are found in most lower crusts. It is interesting to see that the uppermost mantle, the mantle lid, is also rigid and brittle. The reason is the special strength of its ultramafic material, such as peridotite, which consists of 70 percent olivine. It is brittle even up to 800 degrees Celsius. As a consequence, the uppermost mantle shows rupture and earthquakes in tectonically active areas, while the lower crust is generally void of any earthquakes. The transition to the deeper and weak asthenosphere is slow and transient. The age of the crust and the amount of fluids also play a big role in rheology. Older units are cooler, deeper, and more rigid; fluids and fault systems, on the other hand, tend to weaken the crust.

As we look at the rheology of the continental crust, we certainly have to revise any lingering view of a "rigid" lithosphere on top of a ductile asthenosphere. That notion, derived from the oceanic environment, apparently does not hold for the continental crust, and some geologists even call the lower crust "an intracrustal asthenosphere." Fault zones turn sub-horizontal and vanish in the ductile lower crust. Indentations of hard layers into the ductile lower crust are observed, especially in young mountain belts (see Chapter 12), and tectonic escape is based on the ductile lower crust, as will be explained in Chapter 13. In most young areas, tectonic and seismic observations strongly indicate the existence of ductile layers in the lower crust, but they seem to be missing in the old and cold cratons.

Other Tactics

There are, of course, still other geophysical methods that are applied to crust studies. Scientists can observe the gravity field with gravimeters or with satellites that reveal density anomalies. These surveys prepare or complement seismic studies, especially when the goal is to find the undulations of seismic boundaries, or to look for density anomalies, ore bodies, or other resources.

Magnetic surveys have proved very useful in the study of the magnetized ocean floor. Used on land and from airplanes or satellites, magnetometers may also reveal magnetic anomalies caused by volcanic dikes, faults, ore bodies, archeological objects, etc., in the upper crust. Magnetic surveys of the deeper crust, on the other hand, are hampered by the fast decrease of magnetic field strength and by the limited depth of magnetized bodies. Below a Curie point of 500 to 600 degrees Celsius, no strong (ferromagnetic) magnetization is possible. (This temperature range is present in the deep continental crust and in the shallow oceanic mantle.)

Electromagnetic and electric surveys are also used to study the properties of the crust because the fluid-filled layers, graphitic seams, salt layers, or melt pockets are strongly conductive. Geothermal studies often make use of boreholes to measure the temperature, the temperature gradient, and heat conduction of the crust. From the geophysical point of view, temperature plays a dominant role in many processes, but its extrapolation to greater depths can be seriously misleading, mostly because of the circulation of water. Warm or cold circulating waters carry their temperature into the rock unit. This process is very hard to measure from the surface or from shallow boreholes.

Scientists can also use petrological studies to investigate the crust by analyzing samples that were brought to the surface through volcanism or tectonics. These so-called *xenoliths* often provide important information about the material and the pressure and temperature conditions with respect to the crustal depth. Crust samples are also examined in petrophysical laboratories for their elastic and rheological properties. Between some minerals and mineral reactions, there is an equilibrium. This is mainly a function of temperature, sometimes of temperature and pressure. Petrologists analyzing xenoliths or samples from boreholes determine this equilibrium and estimate the temperature (and sometimes the pressure) under which the samples were created. This technique is called *geo-thermometry*, or *geo-thermo-barometry*.

The integration of the various methods, be it on land, in boreholes, or at sea, has revealed the evolution and the architecture of our Earth's crust.

Hydrocarbons

Hydrocarbons are natural compounds with long chains of carbon and hydrogen. Complex combinations form petroleum and natural gas. In nature, hydrocarbons develop from marine microorganisms that are transformed over time and shielded from oxygen in stagnant sediment basins (in so-called *euxinic* environments). Thick, high-pressure coverage or high temperatures can destroy the long chains of hydrocarbons, which is why medium temperature and low to medium depths are necessary for their formation and existence. During a long tectonic history, hydrocarbon-containing ("mother") rocks are usually deformed under the influence of varying tectonic stresses. The light hydrocarbons migrate upwards until they reach a tectonic "trap," where they accumulate (see Figure 11.6). They can be extracted easily if they are in porous sandstone or limestone with some non-permeable rock in the hanging wall. Hydrocarbons are generally separated according to their weight—gas on top and oil below, followed by groundwater at the bottom.

Petroleum Exploration and Exploitation Techniques

As discussed earlier in this chapter, the search for hydrocarbon traps within sediment layers is conducted by methods of reflection seismology. The advancement of computer sciences and global positioning systems led to another great leap forward in these acquisition and processing techniques. For example, the signals received from geo- or hydrophones are now converted into real three-dimensional pictures of the subsurface. Through repeated measurements during the exploitation of an oil field (a so-called *4-D seismic study*), scientists can even detect the migration of oil and the deterioration and modification of an oil field caused by pumping and by intruding water.

New drilling techniques can even push the remaining petroleum into the direction of the production hole by artificially inducing water or carbon dioxide. This technique helps to extract

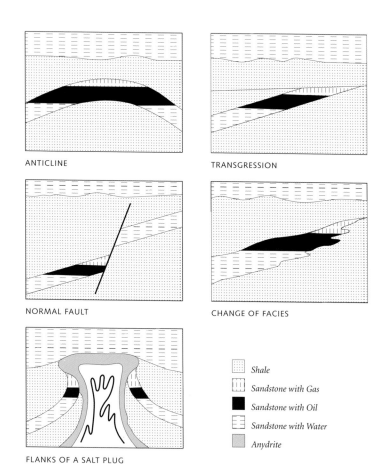

ANTICLINE

TRANSGRESSION

NORMAL FAULT

CHANGE OF FACIES

FLANKS OF A SALT PLUG

- Shale
- Sandstone with Gas
- Sandstone with Oil
- Sandstone with Water
- Anydrite

FIGURE 11.6 Examples of oil traps in sediments (detected by reflection seismology).

the last drops of oil of a given trap—a sophisticated but expensive method. The classical *rotary technique,* practiced for more than 100 years, used a big rotating table on top, a large number of solid pipes, and a hammering or rotating chisel below (often fitted with diamonds). However, used chisels had to be replaced after a few days, and all the pipes had to be extracted (and re-introduced with the new chisel). In addition, the rotary technique could not initiate or control deviations from the vertical.

Real target-oriented flexible drilling became possible only after the development of the *turbine technique.* This method uses a drilling head at depth, rotated by a special kind of muddy solution. Electronic sensors control and change its direction. During the scientific drilling of the 9000-meter-deep KTB hole in southern Germany, for example, special sensors were used that endured temperatures of up to 300 degrees Celsius. At this temperature, some rocks begin to creep and deform, limiting a deeper penetration capability because the hole tends to close after any interruption of drilling. For normal drilling (in colder rock units), the new target-oriented technique can even direct the borehole along horizontal paths or arrange multiple holes from the same drilling platform. Based on new geophysical and geochemical sensors that can "smell" oil, it is possible to perform corrections of deviations during the drilling.

Despite the tremendous improvements in exploration and drilling techniques, the limited oil reserves in partially exploited fields will certainly make exploration and further exploitation increasingly difficult and expensive.

Alternative Sources of Hydrocarbons

There are two more petroleum sources on land that may be explored in the future: oil sands and oil shales. In these rocks, oil is not fluid but has a tar-like viscosity. As a result, it cannot be pumped but has to be extracted by mining techniques. At the time of this writing, several tests are underway in North America and Europe. Exploitation, similar to the technique for shallow lignite, is generally possible but cannot compete economically with low oil prices.

Another source of future hydrocarbon exploration and exploitation are the marine *gas hydrates*. They are locked in so-called *clathrates*, i. e., structures of ice molecules, found in regions of low temperature. Gas hydrates are of white color and were discovered in the deep ocean basins and permafrost shelves of Alaska and Siberia. Efficient exploitation techniques have yet to be developed, but resources are enormous and exceed those of petroleum worldwide.

Natural gas, on the other hand, today is available in much higher quantities than oil. Huge gas fields in Russia, for example, cover the European demand. Combined with methane, it can be converted into a fluid by using steams, heat, and various catalysts. Moreover, natural gas can be extracted by a catalytic synthesis out of coal using a form of the 1920s *Fischer-Tropsch technology*, but competition with the low gas prices is not possible.

Eventually, we might be able to use *biogas* and *biodiesel* made from several plants, usually grain and corn. These can be considered to be renewable sources of energy, but they cannot quantitatively replace our limited oil and gas reserves.

Coal

In contrast to the marine hydrocarbons, coal is created out of swamp vegetation by certain chemical transformations that occur over millions of years. Under increasing pressure, dead trees and plants are first transformed into peat and then into lignite. Even higher tectonic pressure and temperatures are necessary for coal to develop.

The present coal reserves are large and may last for several hundred years. Coal is found in shallow structures in North America and exploited through various surface-oriented mining techniques. Layers in western Europe have determined the industrial structure of large regions for more than 150 years, but all shallow layers are now exploited, and the remaining layers are deep and very expensive to mine.

Lignite, on the other hand, is available in many areas of North America and Europe. The massive dismantling of shallow lignite from huge surface depressions frequently necessitates a re-naturalization of exploited areas. After much hardship regarding the evacuation of people, though, the re-naturalized landscape looks sometimes more "natural" than before.

Other Subsurface-based Resources

Volcanic and fluid-related processes have led to a concentration of various *ore bodies* in the Earth's crust. As mentioned in Chapter 8, very massive banded iron formations (BIF) were created in the Archean and the early Proterozoic. BIF are today the richest iron formations worldwide. Other important ore deposits are those generated from *hydrothermal* (hot water) solutions. There are, for example, the gold deposits in South Dakota; the lead, zinc, and silver deposits in Idaho; and the copper ores in Michigan. All these ores segregated at the bottom of an enormous magma chamber, were later transported into veins and cracks, and deformed under the influence of tectonic stresses. It is interesting to know that the rich copper and iron deposits in the South-American Andes were created hydrothermally in huge magma chambers near active oceanic ridges, from where they were transported to the sides by the drifting oceanic plate, and finally incorporated into the partially magmatic mountains above the subduction zone. The exploration of ores is not easy when underground. They are mostly too small to generate a seismic reflection, and seismologists have to use the scattering properties of these bodies in order to detect them.

Groundwater and *geothermal energy*, too, are exploitable resources that originate in the Earth's subsurface. We will discuss them in the Epilogue to this book.

chapter 12

Formation of Mountains and Basins

When lithospheric plates converge and collide, mountain belts are created. If the plates diverge, rifts, grabens, or basins can be the result. In this chapter, we will discuss the basic types of tectonic processes that create and shape mountains and basins.

Collisions

As far as we know, there are basically three different kinds of collision.

- *Ocean–ocean collisions* are observed primarily in the western Pacific, where the older and heavier Pacific plate subducts below a younger and lighter plate. Along the subduction zone, earthquakes and some basaltic volcanism are the consequences.
- *Ocean–continent collisions* are the most frequent type of collision worldwide. They are most spectacular around the

Pacific, with the exception of California and Oregon (San Andreas strike-slip zone), where the present collision is not perpendicular to the continent but oblique, creating the strike-slip movement along the San Andreas Fault. On the seaward side of an ocean–continent collision, there is first a small bulge with high gravity, followed by a deep-sea trench, and strong earthquakes that follow the subduction zone downward (see Chapter 10). Lava from the subducting and fluid-filled oceanic crust rises to the surface to form the famous "Ring of Fire." While the major part of the elevation comes from compression of the oceanic and the continental plates, the Andes are also formed, in part, by a number of volcanic andesites. Figure 12.1 shows an example of an ocean–continent collision with a rather shallow subduction zone from the Andes (the oceanic Nazca plate is comparatively young). This collision displays an *erosive* (not an accretive) structure. Sediments have been eroded from below and were carried downward on the shoulder of the intruding and subducting oceanic plate. Another special tectonic structure caused by an ocean–continent collision is the *accretional wedge.* It often forms a fore-arc and consists of sediments that are stacked by the approaching oceanic plate.

· *Continent–continent collisions* are the most dramatic tectonic processes. However, they do not produce sharp boundaries. For instance, the collision between the micro-continent India, coming from the south, and Asia started nearly 60 million years ago, and India is still penetrating Asia at a rate of approximately 5 centimeters per year. The compressive power of this collision is distributed over an

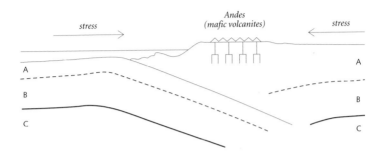

OCEANIC NAZCA PLATE   *SOUTH AMERICAN PLATE*

---   *Moho*
A   *Crust*
B   *Subcrustal Lithosphere*
C   *Asthenosphere*

FIGURE 12.1 A mountain belt is created at a convergent ocean–continent plate boundary. It develops through compressive crustal shortening and intrusions of magmas from below. These magmas rise from the subducted and heated fluid-filled oceanic crust. On their way up, they mix with continental material, producing andesitic lava and explosive volcanism. This formation of folded mountains is called *orogeny*.

area of 2,000 × 3,000 kilometers, with Tibet showing an average elevation of nearly 5000 meters—the largest and widest plateau on Earth. This collision has also produced strong earthquakes in the Himalayas and in Tibet, but nearly all of them are shallow and along prominent thrusts. Only a few of the earthquakes in southern Tibet have occurred at depths greater than 80 kilometers, indicating the intrusion of the cold and rigid Indian subcrustal lithosphere. Studies of the focal plane mechanisms of these shallow earthquakes revealed that they are compressive in a north–south direction in the Himalayas and switch to extension in central Tibet. This strange behavior is documented by geological mapping that shows north–south directed rifts and grabens and east–west strike-slip faults in central Tibet. Rifts and rift valleys, mostly 20 to 30 kilometers wide, are consequences of extensional stresses and deformation more or less perpendicular to the rift axis. A rift trough is termed a *graben*; such troughs have generally been thrown down along faults relative to the rocks on either side. These faults are called *extensional* faults or *normal* faults. Under the oblique attack of stresses, *strike-slip faults* are generated if one crustal block slides past another. They are generally caused by an oblique compression or (seldom) extension. *Thrust faults* or *thrusts* occur during a "head-on" collison. All these tectonic features are characterized by a specific focal plane solution, when active. Under the strong convergence between India and Asia, compressive forces "expel" the middle part of Tibet toward the east with extensional features in the middle and strike-slip faults at the side of the "escaping" middle part. Also, a switch of the direction of seismic anisotropy (explained in Chapter 13) supports the concept that central Tibet is somehow drifting to the east. Such displacements are perpendicular to the compressive forces and are observed in some other young mountains like the Alps or in Turkey. They are called *tectonic extrusion* and *escape.*

Orogeny

The formation of folded mountains is called an *orogeny*, and can easily be understood by imagining a carpet being pushed from two sides. However, this comparison is a bit off because the subsurface of the carpet is generally smooth and even, whereas the subsurface of compressing mountains grows deeper and weaker. The best-investigated compressive orogenic mountain belts are the Alps. Here, the mountains resulted from a multiphase collision between the Adriatica terrane (an African promontory) and Europe. A whole series of deep sedimentary basins in the Jurassic period (see Chapter 14) preceded the onset of collision that reached its maximum some 60 million years ago at the end of the Cretaceous period. In the Alps, there is not much magmatism but strong folding and thrusting, much metamorphism, and a dramatic uplift of the former sediments. Erosion and Pleistocene glaciation shaped the final surface of the Alps.

In the 1960s and 1970s, international seismic refraction surveys determined a crustal root of 55 kilometers depth. Similar root zones were found in nearly all young mountain belts. An exciting picture of current tectonics in the Western Alps was found via reflection surveys carried out by a Swiss consortium in 1989. The

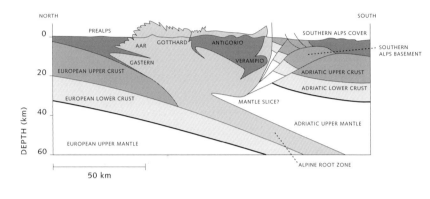

FIGURE 12.2 The Alps, created in a continent–continent collision of the approaching Adriatica and Europe terranes. This is a tectonic interpretation of various Swiss seismic reflection profiles. Note the crustal shortening, thickening, and indentation of many different crustal layers in the upper crust and the massive indentation (interfingering) of the shallow Adriatic mantle into the European crust at depth. The names in the upper part of the figure refer to locations in Switzerland and Italy.

Swiss observed a massive cold and rigid *indenter* penetrating the weak European lower crust from the south. This indenter seems to be a wedge of rigid Adriatic mantle. Such a compressive indentation, or *interfingering*, certainly contributed to the uplift of the Alps and seems to be a general tectonic phenomenon because it is also seen in other young mountain systems (see Figure 12.2).

Old mountain ranges like the European Variscides still show some compressive structures, predominantly in the upper or middle crust. This phenomenon is the result of an enormous collapse that took place at the end of the orogeny about 300 million years ago. During this hot collapse, many granites were formed, mountain roots disappeared, and the former thickened crust was profoundly thinned and modified. Extensional processes caused the strong lamination in the weak lower crust (see, again, Figure 11.4).

Sediment Basins

All the large sediment basins on Earth have been formed by extensional processes and/or subsidence. Early subsidence is caused by a cooling, contraction, and densification of magma that entered a basin early in the extension process. Evidently, extension is a deep process and may open deep faults as pathways for magma. The deepest basin on Earth is the Michigan Basin, with a depth of nearly 30 kilometers and many alternating volcanic and sedimentary layers. Other large basins are found on the Atlantic passive rim around America and Africa/Europe where the old oceanic seafloor got heavy and subsided, but has not yet subducted. Under these circumstances, two types of basins are formed. The first type are basins with intense volcanism. Together with incoming sediments from the adjacent continent they form layers that dip seaward, a phenomenon that has been observed in marine seismic surveys. The second type are rifted margins without volcanism that often include deep fault systems.

Sometimes the formation of two-dimensional rifts turns into an oceanic graben that grows into an oceanic basin like the Atlantic Ocean. Other grabens, like the Red Sea, are halfway between a continental and an oceanic rift. While the Red Sea is still a continental rift in the north, the southern part is already oceanic and shows magnetic stripes dating back some 5 million years. When Pangea began to divide 200 million years ago, many new rifts and basins appeared as new convection patterns formed in the mantle.

Of course, most rifts do not develop into oceans but stop growing after a few million years. The Rio Grande rift in North America and the Rhine-Rhône and the Eger graben in Europe are now passive, although, during some periods in their development, basaltic lava from the mantle came to the surface. Before the opening of the Atlantic, a multitude of rifts appeared between Greenland and Scandinavia but only two remained active and spread. Why only two spreading ridges (and later only one) developed is not known, but it must have something to do with the dimension of the deep convection cell. The East African graben system is presently active, but it might take 1 million more years until "ships will sail to maritime Uganda" (as a poem has it, presented at a 1990 geophysicists' meeting in Washington, D.C.).

In most basins, early rifting can still be observed in seismic reflection images. With slow subsidence, more and more sediments accumulate, and marine saltwater and various organisms are deeply buried, compacted, and transformed to salt layers and hydrocarbons. The subsidence is generally so slow that it is *isostatic*, which means there is an equilibrium between the basin and its surroundings. If there is a different elevation, say, between a basin and its neighborhood, the different weight of the two areas at depth will cause a difference in lithostatic pressure and stresses. In addition, sediments in the basin will cause differences in the pressure and stresses via surrounding crystalline rock. However, there is a tendency, especially in the deep and weak part of the lower crust, to form an equilibrium, and material will creep in until the equilibrium is reached. The compensation is mostly carried out by an *updoming* of the mantle below the basin. Also, a densification of the lower crust, e. g., by the intrusion of dense magmas from the mantle, might sometimes compensate for the light sediments and the lower elevation of the basin. Hence, compensating creep processes will finally establish some sort of isostatic equilibrium at depth.

The very different deformation pattern of basins reflects the large variety of stresses, tectonics, and rheology (warm and young, or old and rigid) that shape them. For example, some smaller basins can develop along strike-slip faults that carry parts of the sediments away and cause some subsidence. These are called "pull-apart basins." There is such a big variety of rifts and grabens that it is hard to find a general concept. We may, however, differentiate between the various ages of basin development, whether it is still in the first phase of extension, in the phase of magma intrusion, or in the phase of magma cooling and subsiding. Temperature-dependent rheology at depth determines the degree and compensation of the accumulating sediments on top and the rise of the heavy mantle below. Rheology also specifies how deep a graben's fault system reaches down, since warm creeping movements at depth destroy these systems. Both rheology and the power of intruding magma determine its lateral spreading. When magma intrudes in various phases, it will first cause a rise and later, when cooling, a fast subsidence and a great depth in the forming basin.

A special topic of basin studies is *basin inversion.* Most of the time, a basin inversion only interrupts the general subsidence for a limited time, but there are some cases where a real orogeny develops. A "normal" basin inversion is caused by a change of the general stress system, i. e., from subsidence-generating *extension* to an uplift-promoting *compression.* The attacking compression might stem from a new active plate boundary in the neighborhood of the basin, such as a conversion with mountain building. We know that tectonic stresses can be transferred over several hundreds of kilometers. To illustrate, the maximum phase of the compression in the Alps, about 60 million years ago, inverted some basins more than 1000 kilometers to the north, all the way up to Scandinavia.

But basin inversion is very selective; it strongly depends on the direction of attacking stresses with respect to the existing fault system and to the quantity of lubricating fluids inside the fault system. Hence, usually only a portion of a given basin is inverted, and normal faults are transformed into thrust faults. The weakness of fault systems, caused by high fluid pressure, and a large accumulation of sediments are the main reasons for a massive inversion. Up to 3000 meters of uplift have been observed in some inverted basins, e. g., near Europe's largest suture, the Tornquist Zone between Sweden and Poland.

Another problem is connected with the question where the stress is transferred. Candidates are the rigid upper crust or the rigid uppermost mantle. Since the rigid upper crust is rather thin and inhomogeneous, the main transfer might use the rigid uppermost mantle. This could explain the large amount of uplifts in some basins, because from the mantle to the sediments a vertical transfer of stresses is necessary, as observed in the inversion of some steep normal faults into vertical thrusts.

We have seen that compressive stresses, related to plate tectonic processes, form the complex shape of mountains with uplift, crustal shortening, indenters, mountain roots, and possible tectonic escape. Extensional stresses initiate the formation of basins, cause subsidence, and favor volcanism. Cooling of intruded magmas supports fast subsidence. The large variety of shapes of extensional structures from rifts, grabens, halfgrabens, and deep and inverted basins, isostatically compensated or not, is the consequence of a complex interplay of viscosity, attacking stresses, and their changes over geologic times. Only careful geophysical and geological studies can reveal the immense complexity of mountain and basin formation.

chapter 13

New Discoveries, New Concepts

Discoveries

New concepts and theories need new descriptions and names. For the geological sciences, the 1970s were the great years of a new paradigm: plate tectonics. Aggressive research and important new discoveries began at this time and have continued into the present. Some discoveries are variations and new aspects of plate tectonics, like the *terrane* concepts or the *delamination* process. Other aspects deal with the observation and theory of the *plume* concept, a special kind of convection, and a process completely separate from plate tectonics. Thanks to major advances in digital seismology, refined computer techniques, and the acquisition of enormous amounts of data, three new methods have been developed in seismology: *seismic tomography, seismic anisotropy,* and a new method of observing seismic boundaries called *receiver function.* All are discussed in this chapter.

Terranes

First, we must consider the concept of terranes. It was discovered that, in addition to the 12 large, established lithospheric plates, there are a number of oceanic or continental microplates, called *terranes*. In the 1970s, David L. Jones of the U.S. Geological Survey studied the Paleozoic and Mesozoic terrane accretion within the cordillera of western North America and established the basic concept of *allochthonous terranes* worldwide. The origins, drifts, and dockings of microplates to continents are important elements of the terrane concept. In both the past and the present, terranes have played a major tectonic role. They originate in an extended magmatic process in the ocean (oceanic terrane) or a rift process near the rim of a continent (continental terrane). Terranes drift with the oceanic plate velocity until they collide with a continent. Then, a head-on collision or a mild accretion takes place (the former often being no different from a continent–continent collision).

Extensive parts of our continents were created by docking terranes. The first discovery of terrane accretion came from Alaska. It was already known that this area was, to a large extent, geologically "exotic." Radioactive dating (Chapter 9) revealed very old formations in the region. In the mountain belts of Alaska, absolute ages of rocks turned out to be very different from and not compatible with adjacent areas. Also, geological and mineralogical data deviated from neighboring strata. Paleomagnetic studies suggested that Alaska had originated in southern geographic latitudes. Today, many different terranes have been found in Alaska, all of which came from the southern Pacific. In the same way, east and southeast Asia and extended areas east of the Ural Mountains have been identified as terranes. Even the path of the Indian subcontinent in the area near Madagascar can be understood as a drift of a terrane (although India is rather large). The whole of Tibet is made up of various terranes that originated in the southern latitudes and accreted to Asia in various phases through the Jurassic and Tertiary periods (see Figure 13.1). Today, many of the islands in the west and southwest Pacific are on their way to southeast Asia, where they will arrive in about 50 to 100 million years.

The biggest surprise of recent times came from the search for terranes in Europe. In the Proterozoic, there was a terrane accumulation in the southwest part of the old Baltica craton, as clearly observed by marine reflection seismology. Around 400 million years ago, when the great collision between the continents Laurasia (North America and Greenland) and Baltica (or Fennoscandia) took place, a terrane from the south (East Avalonia) joined the collision and created the long Caledonian Deformation Zone (CDF). At the same time, West Avalonia joined Laurasia, which today is located near Newfoundland. These terranes were continental. They were rifted apart from the northern rim of Gondwana (present-day Africa) and followed the drift of the Paleo-Tethys ocean to the north until they docked on a continent. The CDF can still be seen by marine reflection surveys from the North Sea to the Baltic Sea. It shows an indentation pattern at depth similar to that found in the Alps (see Chapter 12). This tectonic evolution could be said to be fortuitous in the sense that no additional collision or collapse modified the area later on.

At first, classical plate tectonics could not explain the origin of the Variscan Mountains in western and central Europe because the early studies did not detect signs of any oceans between the various mountain ranges. But the terrane concept, including the

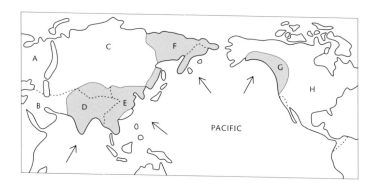

A   *Europe*
B   *Arabia*
C   *Siberia*
D   *India*
E   *Sino-korea*
F   *Kolyma*
G   *Alaska*
H   *America*

FIGURE 13.1 Terranes in the India-Pacific area.

assumption that there were only small oceans between the terranes, did explain the tectonics, and eventually traces of some oceans were found in both England and Germany. The terranes were continental and were (again) split apart from the northern rim of Gondwana, presumably by a large rift process that was possibly caused by a *plume*, and then drifted north. They accreted to northern Europe, mainly to the older Avalonia terrane, between 380 to 300 million years ago. They also helped form various high mountain chains, including the Variscides, which then extended from today's Appalachians to the Iberian Peninsula, to Ireland, and to Poland in the east.

Shortly after this orogeny, the stress system changed from compression to strike-slip and later to extension, causing the huge mountains to collapse. This collapse was connected with a *delamination* process that produced a huge amount of heat, a large amount of granitic rocks, and a complete transformation of the lower crust, which then became thin and laminated (see, again, Figure 11.5). This process concentrated at the center of the Variscides, while further outside the southern parts of Avalonia were overthrust and reworked.

The young Alpine chains in southern Europe were created by a rather recent terrane accretion. The Alps were compressed by the Adriatica terrane, a promontory of Africa that drifted north. Crustal shortening and thickening took place, strata were folded and uplifted, huge crustal roots developed (more than 55 kilometers deep), and there were strong indentations of the hard and rigid Adriatic mantle into the weak European crust (as shown in Figure 12.2). Eventually, parts of the eastern and southern Alps, under the influence of compressive tectonic forces from the south, escaped to the east toward the weak Pannonian basin in present-day Hungary. It must be pointed out that such recognition and analysis of a terrane could only have been achieved through the close interplay of geophysics (mainly by studying reflection profiles) with age dating and careful geological mapping of exotic rocks and structures.

Delamination

Another new and important concept of plate tectonics is *delamination.* Delamination helps us understand instability in a subcrustal, mainly continental, lithosphere, which shows similar effects as an oceanic subduction in its development of negative buoyancy. A warm, weak, and thick lower crust acts as an ideal decoupling layer, and a phase transition from gabbro to eclogite facies at the bottom of a thick crust may provide the main negative buoyancy. In contrast to an oceanic subduction, only the lowermost crust and the upper mantle become unstable and sink. The light upper crust generally is not involved in a delamination, and only very strong compressive sutures are able to pull some of the upper crust down. Figure 13.2 shows three examples of a beginning delamination.

Another consequence of delamination is the mobilization of the asthenosphere after it has been penetrated by the cold, downward-moving subcrustal lithosphere. Hot, asthenopheric material rises and causes volcanism and strong heating and weakening of the lithosphere, especially of the lower crust, where the formation of granite is initiated. This happened during the collapse of the Variscides, where the over-thickened crust of the mountain ranges initiated the delamination.

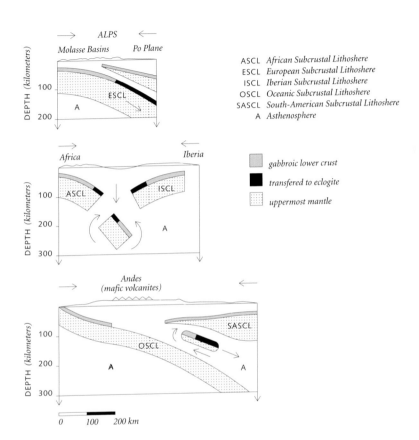

FIGURE 13.2 Three examples of the first phase of delamination. A: The European continental subcrustal lithosphere and some material of the lower crust are delaminated (guided) into the deeper mantle. B: A piece of lithosphere between Africa and Iberia is being delaminated, moving to greater depth. C: A small piece of lithosphere above the subduction zone has been delaminated and is on its way down.

Plumes

Heating and volcanism may also be caused by a *plume*. The plume concept is independent from plate tectonics and is based on observations of volcanic patterns on the surface (also called *hot spots*) and on the mapping of the Earth's mantle by *seismic tomography*. The latter has revealed special anomalies of seismic velocities that are interpreted as temperature anomalies. In the case of Iceland, a plume seems to have created a divergent plate boundary called the North Atlantic Ridge. However, all the other plumes on Earth are accidentally distributed in oceans and continents.

The Hawaiian island chain is especially impressive in this respect. The chain is nearly 7000 kilometers long, and at least 90 volcanoes were active there over the last 70 million years. The oldest was in the northwest, while the younger volcanoes were in the southeast. As mentioned in Chapter 10, the Hawaiian chain has emerged because the Pacific plate migrates at a rate of 10 centimeters per year over a rather stable plume. (The Hawaiian plume has existed for nearly 100 million years and is one of the oldest of present-day plumes.) As a consequence, all volcanoes get younger toward the southeast, and the very youngest, the small "child" of Kilauea on the big island of Hawaii, is still in its very infancy, its top today being some 1000 meters below sea level. A strong bend in the Hawaiian chain can be observed at an age of 70 million years, possibly connected to a re-orientation of the Pacific plate that occurred when India collided with Asia.

In the past, plumes have established large basaltic provinces in oceans and continents. The expelled gases and soot particles of strong volcanism also seem to be responsible for cooling the atmosphere and the climate, and possibly even for biological extinction. An especially strong plume activity, a *super-plume*,

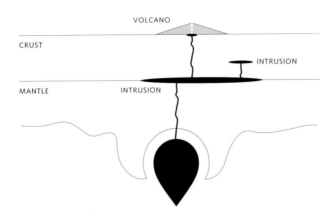

FIGURE 13.3 A plume on its way upward. Its heat has already melted the rocks in the lower crust and generated volcanism. The signatures around the plume show the assumed influence of the rising plume on its surrounding.

existed in the late Cretaceous, about 120 to 65 million years ago, and covered the surface of the whole western Pacific with numerous volcanoes. At this point, it is important to note two additional exciting observations. First, the magnetic field in the late Cretaceous did not change its polarity for nearly 60 million years. If this super-plume really originated at the core–mantle boundary, it might have influenced even the temperature gradient in the fluid core and caused the geodynamo to rotate more steadily. Another observation is the deterioration of the climate in the late Cretaceous, possibly caused by the super-plume that, in turn, might have affected the lives of many biological species, including the dinosaurs. At the present time, some indication of a super-plume below Africa has been found.

How do plumes originate? It is generally assumed that plumes come from great depths. The magma carries many incompatible elements—trace elements like rubidium, thorium, uranium, and barium that are surmised to exist in the lower mantle. Most researchers agree that plumes must originate at a "thermal boundary layer," such as the boundary between the upper and the lower mantle at a depth of about 660 kilometers, or the core–mantle boundary at a depth of 3000 kilometers. Seismic tomography suggests that all large plumes originate at the core–mantle boundary, the so-called D"-layer (see Chapter 4). This highly complex boundary has, as it turned out, a variable thickness of 100 to 200 kilometers. Its temperature ranges between 3000 and 4000 degrees Celsius.

The plume concept also seems to solve a very old controversy about convection patterns in the mantle. There were arguments for a *whole-mantle convection* and for separate convection patterns in the upper and in the lower mantle, separated by the well-known

660-kilometer discontinuity where spinel transforms to perovskite (see Chapter 4). Data from seismic tomography suggests a compromise: strong plumes and strong subduction zones may penetrate the discontinuity between the upper and the lower mantles, otherwise the discontinuity would provide a separating boundary.

Cooling and contracting plumes also appear to govern the first stages of basin formation by strong subsidence. Lots of plumes are observed on the young volcanic surface of the hot planet Venus. There, cooling and contracting plumes are held responsible for a cyclic resurfacing process, a kind of delamination that may even carry part of the surface down.

New Seismological Concepts

The science of seismology also developed new methods over the past 30 years. As mentioned in Chapter 7, an important area of study is the propagation of P- and S-waves with different speed in different directions in practically all layers in the crust and the upper mantle, a phenomenon called *seismic anisotropy*. Research in this field is made possible by a tremendous number of receivers and continuing investigations of earthquakes of all magnitudes. There are various types of seismic anisotropy, but the most important one is *directional*, or *azimuthal*, *anisotropy*, which is based on the lattice-preferred orientation of anisotropic minerals. Since the dominant mineral in the upper mantle is the orthorhombic mineral olivine, its orientation in the creep direction produces strong anisotropies. Creep in the hot mantle is the cause for anisotropies in oceans and continents. Recently, it was discovered that young mountain belts develop an anisotropy parallel to their structural axis. This is explained by a degree of escape movement that is influenced by compressive forces perpendicular to an elevated mountain belt.

Another new seismological technique for the study of the crust and the upper mantle is the *receiver function method.* Broadband seismological stations along a profile make use of the difference of refracted arrival times between P- and S-waves. The reason for this difference is that incoming P-waves split into P- and S-waves at a seismic boundary. Both waves, preferably those caused by teleseismic events, arriving at steep angles, are recorded and stacked. In addition to the P- and S-signals from the weak intracrustal boundaries, signals received from the Moho and from the 410-kilometer (olivine–spinel) and 660-kilometer (spinel–perovskite) discontinuities show travel time differences. Even the polarity of the "generated" S-waves can be displayed. Many tectonic problems may yet be solved by this inexpensive method, though the frequencies are rather low and limiting. The method is still in the phase of testing but provides a great opportunity for supporting other seismic methods quasi-independently.

chapter 14

The Phanerozoic: The Last 600 Million Years

Just before a rapid diversification of marine life began, culminating in the Cambrian Explosion between 525 and 575 million years ago, an extreme glaciation occurred. As mentioned in Chapter 8, the mega-continent Rodinia dispersed in the Neo-Proterozoic, around 770 million years ago, and left many small continents scattered near the equator. Increased rainfall in the tropics "scrubbed" the atmosphere, absorbing carbon dioxide out of the air, increasing erosion, and reducing global temperatures. Large ice packs formed in the polar oceans, and a positive feedback mechanism started. The white ice reflected more solar energy than the darker seawater, and this in turn drove temperatures even lower. Finally, the whole planet was covered by ice; temperatures dropped to –50 degrees Celsius, creating an effect called *snowball Earth*, which lasted a few million years.

Without rainfall, carbon dioxide expelled from volcanoes gradually accumulated, the planet warmed, and the sea ice thinned. Open water was re-established in the tropics and began absorbing solar energy. This, in turn, created yet another feedback mechanism, this time toward global warming. A hot, wet Earth washed bicarbonate and other ions into the oceans, where they formed carbonate sediments. New life forms with multi-cellular structures developed, and the first hard skeletons appeared after a prolonged genetic isolation.

The Evolution of Life

Based on the strong diversification of marine life in the Cambrian, a biostratigraphic time scale was established at the end of the nineteenth century. It was a relative time scale based on the fossils in alternating strata on top of each other. The preferred subjects for investigation were non-deformed and horizontally deposited sediments. The new science of paleontology dealt with individual fossils and certain groups of fossils, their variations over time, and extinctions. A new branch, *biostratigraphy*, specialized in marine strata and was later augmented by seismic and magnetic stratigraphy.

A theory of the evolution of life was created by joint research in geology, paleontology, and biology. Through this collaborative study, scientists learned that life did not develop smoothly and continuously, but was often interrupted by extinctions. Such sudden changes in life forms had guided paleontologists to formulate, at the end of the nineteenth century, the first relative geologic time-scale.

The largest extinction took place toward the end of the Paleozoic, when 90 percent of all life on Earth was eliminated. Another major extinction occurred at the end of the Mesozoic, when 50 percent of the then-existing species, including the dinosaurs, disappeared. The latter event was almost certainly caused by the impact of a giant meteorite. This theory was supported by the discovery of the 200-kilometer-wide Chicxulub impact crater on Mexico's Yucatan peninsula, the place where scientists found remnants of a huge meteorite that had hit partly a land area and partly the sea. Over the last 65 million years, about 400 meters of thick carbonates and limestone were deposited, hiding the crater. The site was found nearly 20 years after Luis Alvarez of the University of California at Berkeley had postulated an enormous impact, based on iridium anomalies worldwide. Marine seismic surveys and gravity measurements on land confirmed the existence of the crater and the existence of much impact material.

The consequences of the impact were severe. Tremendous flooding occurred, and the expelled dust and gas in the atmosphere practically obliterated sunlight for several hundred years. As a result, widespread extinction occurred, even in the oceans, and only a few life forms, such as frogs, survived. Some researchers think massive volcanism triggered by plumes at about the same time on the Voering Plateau in Scandinavia and the Deccan Traps in India aggravated the extinction. But new life forms soon developed, especially diversified mammals (see also Chapter 15). In addition to the two big extinctions mentioned above, a number of minor extinctions took place during the Phanerozoic, and all of them seem to correlate with the impacts of smaller meteorites.

ERA	PERIOD	EPOCH	YEARS BEFORE PRESENT
Paleozoic	Cambrian		570–505 million
	Ordovician		505–438 million
	Silurian		438–408 million
	Devonian		408–360 million
	Carboniferous		360–286 million
	Permian		286–245 million
Mesozoic	Triassic		245–208 million
	Jurassic		208–144 million
	Cretaceous		144–65 million
Cenozoic	Tertiary	Paleocene	65–57 million
		Eocene	57–37 million
		Oligocene	37–24 million
		Miocene	24–5.3 million
		Pliocene	5.3–1.6 million
Cenozoic	Quaternary	Pleistocene	1.6 million–10,000
		Holocene	10,000 to present

TABLE 14.1. Eras, Periods, and Epochs of the Phanerozoic Age

The Eras of the Phanerozoic

After the so-called Archean and Proterozoic *ages*, there was a third age, called the Phanerozoic, which itself is divided into the three *eras* Paleozoic, Mesozoic, and Cenozoic (see Table 14.1).

The Paleozoic

The first period in the Paleozoic is the Cambrian (570 to 505 million years ago). Earth's climate was warm, and there were still many small continents around the equator. At this time, an explosion of life forms with hard shells and skeletons took place. Trilobites, the forerunners of crayfish, evolved at this time (Figure 14.1). They consisted of three pieces and ultimately developed into 600 different species. The first shell-like brachiopodes and mollusks (forerunners of snails) also appeared at this time.

The *Ordovician* period, between 510 and 438 million years ago, saw a great increase in volcanism. From the southern continent Gondvana, the terranes Avalonia and Armorika were rifted apart and drifted to the north. Then, the Laurasia and Baltica cratons drew near to each other as Avalonia arrived from the south. Limestone and carbonates developed, and the number of trilobite species decreased in conjunction with the evolution of the brachiopodes and mollusks.

During the *Silurian* period, 438 to 408 million years ago, the collision between Laurasia, Baltica, and the Avalonia terrane occurred in the north. West-Avalonia accreted to Laurentia (Greenland and northern North America, mainly the northern Appalachians, Nova Scotia, and Newfoundland), while East-Avalonia docked to Baltica, forming the Norwegian-Scottish

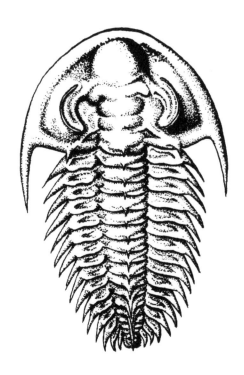

FIGURE 14.1. A trilobite class of anthropods
that evolved during the Cambrian period.

Mountains. Figure 14.2 shows the age provinces in North America with the docking of West-Avalonia in the northeast. Toward the end of the Silurian, the proto-Atlantic, called the *Iapetus Ocean*, began to shrink. Many marine life forms began to move onto the land, sheltered by an ozone layer on top of a nitrogen-oxygen atmosphere, not too different from today's atmosphere. Many microorganisms on the oceanic shelf began to form layers of hydrocarbons, and some trilobites were still present. Around the same time, the first fish developed in the oceans, along with reefs and corals. From examining the fossil remains of ancient corals, scientists have concluded that the tides were strong, that the Moon was closer, and that it took the Earth some 410 days to orbit the Sun.

During the *Devonian* period, between 408 and 360 million years ago, the Caledonian Mountains developed further, and the Iapetus disappeared completely. Additional terranes split apart from the approaching Gondwana continent and drifted north. In North America, the massive Michigan Basin developed. The first marine mammals, radiolarians with rigid skeletons, appeared, and fish diversified. On land, early forests began to fill with ferns and algae.

During the *Carboniferous* period, between 360 and 286 million years ago, an enormous mountain range, the Variscides, formed in a collision of terranes from the south (see also Chapter 12). Toward the end of the Carboniferous, rifts, grabens, and huge strike-slip faults initiated the collapse of the mountain ranges, including the Appalachians. Only the Ural Mountains continued building, due to the arrival of terranes from the east. The climate was humid and warm, and layers of coal were deposited in swamps worldwide. A further diversification of land flora took place, and the first conifers appeared.

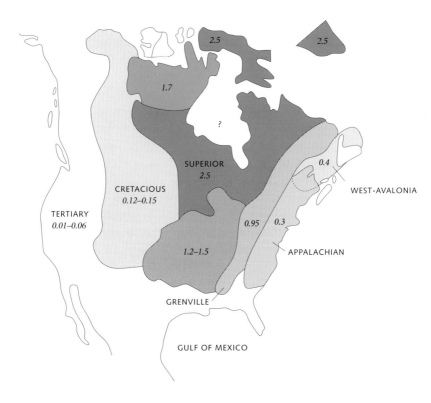

FIGURE 14.2. Age provinces in North America. Note that the boundaries are approximate, sometimes overlapping or covered by younger sediments. Numbers indicate the age (million years ago).

Then, in the *Permian*, some 286 to 250 million years ago, the huge continent Pangea was created, comprising nearly all the continents of the present Earth. The climate became warm and dry. Large portions of polar ice began to melt, the sea level rose, and a huge transition transferred seawater onto the lowlands of continents, which eventually resulted in the formation of land-based reservoirs of hydrocarbons based on marine microorganisms. On the western rim of South America, an ocean–continent collision triggered the formation of the Andes. The collapse of the Variscan Mountains continued. Flora and fauna diversified again, while large foraminiferae (a form of marine protozoans) and marine sea urchins developed. Reptiles diversified. Some even survived the great extinction that at the end of the Permian killed nearly 90 percent of all life forms, including the last trilobites and the early corals.

The Mesozoic

The *Trias*, 250 to 206 million years ago, was the first period of the oncoming Mesozoic. The super-continent Pangea was still intact and slowly moving northward. The climate became humid and warm again. Continental rifts and a strong volcanism developed, heralding the eventual breakup of Pangea. As mentioned, the huge extinction at the end of the Paleozoic eliminated all but 10 percent of the then-living species. However, new species were able to exploit the vacant niches, and by the end of the Trias, the first birds and gliding (or possibly even flying) reptiles had appeared. They diversified, some developing into the first dinosaurs. Various shells, the horsetail plant, and more forms of conifers also appeared at this time.

During the *Jurassic*, between 206 and 144 million years ago, Pangea began dividing. In the south, the Atlantic opened, soon followed by an opening in the Caribbean. The large Tethys Ocean south of Europe emerged, dividing Africa from the northern continents. In South America, the Andes continued to grow. Large shelf areas developed, providing good conditions for the development of the spineless ammonites (Figure 14.3), which eventually split into some 8000 species and are used to characterize certain locations and geological sections. Marine transgressions created more hydrocarbons in the flooded lowlands. The terrestrial flora began to resemble today's species. Conifers changed again, and the first real bird, the archeopteryx, emerged. (A well-preserved archeopteryx skeleton was found in the shales of Sollnhofen, Germany.) Flying dinosaurs appeared, as did crocodiles, the latter in a form similar to today's.

During the *Cretaceous* period, 144 to 65 million years ago, strong seafloor spreading caused a significant increase in the area of the Atlantic Ocean. In the western Pacific, a kind of *mega-plume* created numerous volcanic spots. As a result, large amounts of carbon dioxide entered the atmosphere, causing a greenhouse effect. In this time period, the early and middle Cretaceous, temperature and carbon dioxide content in the atmosphere reached a maximum. The polar ice caps melted, and the sea level rose, leading to another geological transition worldwide. At the same time, many oil- and gas-containing layers were created out of marine microorganisms and plankton. Toward the end of the Cretaceous and into the Mesozoic, the Rocky Mountains and the Alps began to form. The climate deteriorated again, it got colder, ice caps formed, and many animals suffered from the lower temperatures, which decreased the food supply. Oaks, willows, and other trees of the present appeared. Plants with pollens, real

FIGURE 14.3. An example of the ammonite
group of mollusks that evolved in the
Jurassic period.

insects, and warm-blooded animals were present. Many of the new mammals and the last of the dinosaurs may have competed with each other.

The Cenozoic

In the beginning of the *Tertiary*, 65 to 1.6 million years ago, the second great extinction of the Phanerozoic took place, triggered by the impact of a huge meteorite as described earlier in this chapter. Flooding, darkness, a fall in sea level, and cooling (conditions predicted today for a hypothetical nuclear winter) were the consequences. The North Atlantic and other northern oceans developed. Australia separated from Antarctica. Between 60 and 45 million years ago, India collided with Asia, triggering the formation of the Himalayas. At the same time, the Hawaiian chain bent, as the whole Pacific plate changed direction. In the middle of the Tertiary, early horses, pigs, giraffes, camels, and the first primates appeared, but most were smaller than today's species. Between 5 and 6 million years ago, the earliest evidence of hominids begins to appear in Africa's archeological record. In Australia, which was now isolated, marsupials developed. Antarctica was isolated, too, and began drifting toward the South Pole. The low temperatures around the South Pole decreased even more through a marine circular current around Antarctica, where all land animals were doomed to perish. The world climate deteriorated, and polar glaciers became more frequent, making rapid advances.

The *Quaternary* with the *Pleistocene*, about 1.6 million to 10,000 years ago, and the subsequently following *Holocene* were the time of human evolution in Africa. During the last ice age, possibly as early as 30,000 years ago, humans reached America via a land bridge between Siberia and Alaska. A great climatic change occurred at the beginning of the Holocene, as a long ice age evolved into a warm climate. This warming up seems to have been connected with a worldwide change of ocean currents, but why the ocean currents changed is by no means clear. In the long term, the accretion of India to Asia and, later, the amalgamation of South and North America blocked worldwide east–west currents in lower latitudes, and this could have been the major cause. On the other hand, minor changes in seawater temperature might have initiated a big change in ocean circulation, a so-called *self-amplifying* process—a process that is also not well understood. The development of hominids during this period in which Earth's surface environment frequently changed between colder and warmer temperatures is described in the next chapter.

Biological Evolution

This chapter continues the description of the complex and inter-dependent influences of biological and geological processes in the development of Earth. Many of these processes were initiated by pure chance, and proceeded not smoothly, but by fits and starts. The first organic bonds may even have emerged during the formative phase but were "disinfected" again by huge meteorite impacts. Whether those bonds developed again in warm ponds of water or came via comets or other impacting bodies is still hotly debated.

A Product of Chance

When the period of intense cosmic bombardment ended at the beginning of the Archean, about 3.9 billion years ago, the young Sun only had 70 percent of its current luminosity. The Earth rotated in about 15 hours, and the nearby Moon generated tremendous tides. The Earth lacked both oxygen and an ozone layer, and our carbon dioxide (and some nitrogen) atmosphere was not much different from that of our neighbor planets Mars and Venus. Through a very fortunate coincidence, the Earth had assumed a favorable distance to the Sun and a nearly circular orbit, conditions that moderated both the planet's radiation and its temperature. By chance, Earth had the right size to hold some critical gases in certain concentrations in the atmosphere and exhale other gases (water and carbon dioxide) from its volcanoes.

Calculations show that if the Earth had been 6 percent closer to the Sun, a mere 9 million kilometers out of a total of 150 million, it would have developed an early *greenhouse* effect, as on Venus; a certain larger distance would have resulted in massive glaciation. A smaller-sized Earth with less gravity would not have held gases with a medium atomic weight (like oxygen) in the atmosphere, and we would have shared the fate of tiny Mars, which only has 38 percent of the Earth's surface gravity and does not have nitrogen or oxygen in its atmosphere. In addition, the internal heat would have "escaped," thus limiting the duration of any plate tectonics-related activity. A body larger than the Earth would have hampered the release of gases from the interior and attracted lighter gases such as helium or hydrogen, thus creating an environment hostile to life as we know it. The moderate temperature in the atmosphere permitted gaseous water to condense into fluid water, and allowed an early form of rain to wash out the carbon dioxide and accumulate in ponds and seas. (We have to remember that, under our current atmospheric pressure, fluid water can only form in the very small temperature range between 0 and 100 degrees Celsius.) In addition, the nearly circular orbit of our Earth around the Sun and a moderate and nearly constant inclination of the Earth's axis (approximately 23 degrees) allow only moderate variations in solar radiation and climate. Certainly, several chance occurances combined to create favorable conditions for the development of life on Earth.

Early Life Hypotheses

In terms of biological development, perhaps the most critical developments in the environment of the young Earth were those that allowed for the formation of fluid water. The first primitive life probably developed in warm ponds, although a small group of researchers still believe that it came from comets or other celestial bodies because various amino acids (forerunners of proteins) have been found in the ice structure of some comets. This extra-terrestrial hypothesis was recently supported by the analysis of the ALH 84001 meteorite from Mars, which seemed to contain traces of bacteria.

Another hypothesis about the origins of life on our planet is based on the recent observation of highly diverse, structurally simple ecosystems thriving at the superheated vents on the seafloor. There, living creatures exist in isolation from sunlight, drawing energy from the Earth's internal heat. Hot, mineral-laden water rises below the seafloor, and various minerals seem to act as templates, scaffolds, and catalysts for the development of primitive life.

Whatever caused the origin of the first primitive organisms, they needed shelter from the harsh ultraviolet radiation from the Sun. These early life forms also required light or thermal energy to develop photosynthesis in order to produce sugar and oxygen out of carbon dioxide. These conditions are found in water at a depth of 5 to 15 meters, but they also appear to be present in deep hydrothermal vents. Whether lightning on the surface of the Earth helped to form complex organic molecules (as found in some classical experiments) is an open question. In addition, scientists still don't know how the first proteins with large chains of amino acids formed. However, one assumes that the crystal growth of certain minerals in clays, sulfides, or iron supported the first critical self-replications. This ability to replicate is considered the beginning of life. Accidental errors during the copying of the genetic information cause variation and mutations. The selection of the most able organisms under the given environment—Darwin's idea of the *natural selection*—finally organizes the "tree of life."

Although life on Earth started rather early, it was restricted to microscopic organisms for about 3 billion years. In our understanding, life is a stable chemical system with the capacity to evolve with the help of three components: *reproduction, mutation,* and *selection*. It took an enormous amount of time (from 4 to 1 billion years ago) for the first multicellular life forms to evolve. In the beginning, life had not yet developed the genetic DNA code but used RNA chains, which can initiate certain replications. Between 3 and 4 billion years ago, some organisms developed a shelter, or membrane, resulting in a structure that can be regarded as the first cell. Out of this, a three-part "family tree" developed based on *bacteria, archaea* (with one cell only), and *eukaryotes* (a cell with a membrane and a few chromosomes). This last group eventually developed multiple-cell systems which, in turn, led to the evolution of animals and humans. Figure 15.1 shows the "tree of life" with the three main life forms and several minor creations.

Even the simplest organisms, shown at the bottom of Figure 15.1, had already developed the chemical and biological condi-

EUKARYOTES

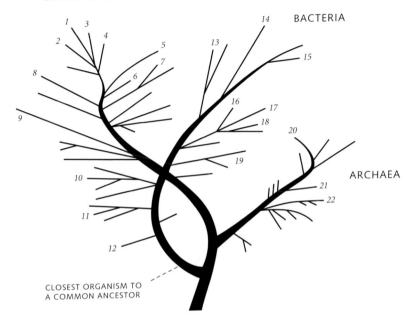

1 ANIMALS (*including humans*)	*12* AQUIFICALES
2 PLANTS	*13* SPIROCHETES
3 FUNGI	*14* GREEN SULFUR BACTERIA
4 STRAMENOPILES	*15* CELLULOSE-DIGESTING BACTERIA
5 ALVEOLATES	*16* AGROBACTERIA
6 RED ALGAE	*17* MITOCHONDRIA
7 ENTAMOEBAE	*18* E. COLI
8 SLIME MOLDS	*19* CYANOBACTERIA AND CHLOROPLASTS
9 MICROSPORIDIANS	*20* SALT-LOVING ARCHAEA
10 GRAM-POSITIVE BACTERIA	*21* METHANE-LOVING ARCHAEA
11 GREEN NONSULFUR BACTERIA	*22* ACID-LOVING ARCHAEA

FIGURE 15.1 The "tree of life" with its three branches of organisms from thermophiles (heat-loving) to archaea, bacteria, and eukaryotes. Multicellular life, including animals and humans, developed from the eukaryote line.

tions for self-replication and a DNA code. The use of the same building blocks and structure for DNA from the most primitive bacteria up to multicellular animals and humans supports the assumption that life started by chance at a particular time in one location. It is possible that there were other competitors, but a selection of DNA structure and replication started early and became global.

The diversification of life forms accelerated after life had created sheltering cells. Toward the end of the Archean, about 2.5 billion years ago, the banded iron formations (BIF) had used nearly all of the available oxygen that had been produced in the sea by photosynthesis. Now, the oxygen could migrate—even into the atmosphere. Oxygen was aggressive, killing many competing microbes that lacked a cell structure, and only organisms with cells survived. These cells later evolved the capacity to take advantage of the available oxygen. At that time, more and more carbon dioxide was washed out of the atmosphere and transformed into carbonates on the seafloor. Argon formed as potassium decayed and increased to about 1 percent, the amount that is found in today's atmosphere.

Multicellular Life

Some organisms with more cells have been sporadically observed in the early Proterozoic, around 2 billion years ago. Between 1 and 1.3 billion years ago, a "sexual revolution" began. By mixing genes of two different individuals (as opposed to self-replication), organisms had more chances for mutation and selection. Some biologists call this "the first explosion of life," but life still remained microscopic. As mentioned in Chapter 14, the real "explosion" occurred at the beginning of the Cambrian, after the "snowball Earth" had turned into a "hothouse." To study the development of hominids, we jump to the Tertiary period, about 65 to 1.6 million years ago.

Starting approximately 60 million years ago, a massive diversification of mammals began. About 50 million years ago, ape-like animals begin to appear in the paleontological and archeological record. These early primates divided into gibbons around 22 million years ago, orangutans about 17 million years ago, gorillas about 10 million years ago, and chimpanzees about 7 million years ago. Between 7 and 5 million years ago, some species became bipedal, walking almost fully upright. Based on some fragmentary fossils, the oldest known hominid is *Ardipithecus ramidus*, which first appeared about 4.4 million years ago. In Ethiopia's Afar region, several remnants of *Homo afarensis* have been found that are nearly 4 million years old.

The whole development of humans is described in simplified form in Figure 15.2. Most surprising is the fast development of the human brain, which started to increase in size about 2 million years ago and, within 2 million years, had become three times larger. Some scientists assume a "hardware-software" feedback system, including the development of language and the larynx. According to this hypothesis, the "hardware," including the brain and its complex cerebral cortex, enlarged and diversified rapidly because its function was continuously challenged by sensory impulses from the peripheral nervous system. This, in turn, caused complex learning and thinking—the "software"—to progress. A strong positive feedback loop led to a steady enhancement of both the "hardware" and the "software." It is interesting

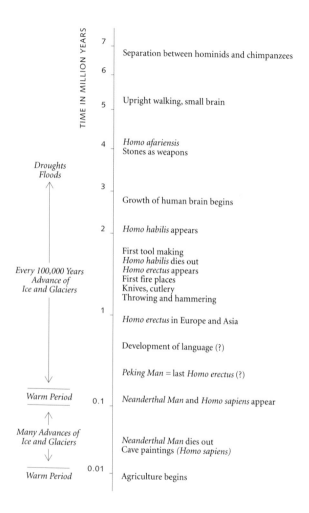

FIGURE 15.2 The evolution of hominids and the corresponding temperature fluctuations.

that, in the development of the hominids, several parallel lines—similar to those of many animal species—are also found. The most recent parallel development is that of *Homo neanderthalensis* and *Homo sapiens*, beginning about 100,000 years ago. They seemed to have lived nearly side by side for at least 70,000 years. Neanderthals had a different head shape and a shorter larynx. Although they excelled as skillful toolmakers, their language capacity was probably limited. It was the development of a refined language, a capacity closely connected with complex and abstract thinking, that most likely gave *Homo sapiens* a certain advantage. Whether the two species "interbred," or what other cause might have induced the disappearance of the Neanderthals, is still a matter of debate.

Human Evolution and Climate

As is shown in Figure 15.2, there seems to be a close connection between the development of humans and Earth's climate. The glaciers began advancing about 2.5 million years ago, and since that time, the climate has generally deteriorated with periodic changes between ice ages and warmer periods. The explanation for periodic changes was laid out by the Serbian astrophysicist Milutin Milankovitch in the 1920s. He noticed that the Earth's axis wobbled slightly and did not maintain a constant inclination. Milankovitch also observed that the elliptic orbit of our planet around the Sun had periodic undulations and could change its ellipticity slightly. It was already known that these periodic irregularities are caused by the slightly varying gravitational pull of neighboring planets, especially Venus and Jupiter. As a consequence, solar radiation varied periodically—at around 20,000, 41,000, and 100,000 years. The *Milankovitch Cycle* accurately describes the amount of solar energy received per unit of time at various latitudes.

Forty to fifty years after it had been calculated, the Milankovitch Cycle was verified by temperature studies in bore-holes in the Pacific, Greenland, and Antarctica. However, in nature, its effects are mostly hidden, and only quiet sedimentation in arid countries show the periodic cycle. For example, escarpments in the form of alternating horizontal bands of various sediments that were definitely created under different climatic conditions reveal time periods that correspond to the Milankovitch Cycle. In the normal temperature display of the last 110,000 years, the cycle is often obscured (as shown Figure 15.3) because of a nonlinear resonance of the climate system and because some feedback mechanisms took place, especially in the global *carbon cycle*. The carbon cycle is a biochemical cycle in which carbon is transferred from air to soil, then to plant and animal life, and back to soil or sea. The amount of carbon generally depends on temperature, but it also might increase or decrease the actual life forms and the temperature again. This "feedback" system is very complex and delays or modifies the Milankovitch Cycle, which often only acts as a trigger of climatic variations. This, in turn, causes asymmetries in the temperature trend, leading to a fast transition from cold to warm (again, see Figure 15.3) and a slow change from warm to cold. In addition, volcanism, cloud formation, and changes of ocean currents complicate the picture. For instance, the Gulf Stream, which apparently assumed its south–north direction around 10,000 years ago, is considered to

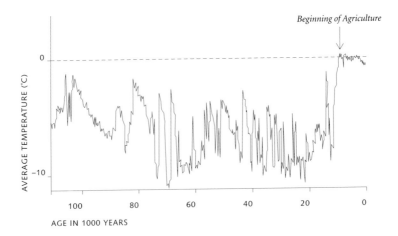

FIGURE 15.3 Ice cores from Greenland
reveal the average temperature in the last
110,000 years. The ratio $^{18}O/^{16}O$ is
determined by an analysis of air bubbles in
the cores. The lower the temperature, the
smaller the ^{18}O contribution. Note the
rapid change 10,000 years ago.

have contributed significantly to the change from the last ice age to the present warm period.

At the same time as the large advances of ice started, about 2.5 million years ago, the human brain began to grow rapidly. The first human fireplaces were dated to 1.5 million years ago, though tool-making seems to have been a bit slow during that time. Strong ice advances appeared 800,000, 460,000, and 230,000 years ago and caused the migration of humans. The development of temperature can be mapped accurately for the last 110,000 years on the basis of air bubbles found in ice cores, age determination using the carbon dating method, and temperature measurements through the relation of the stable oxygen isotopes ^{16}O and ^{18}O.

It appears that the correlation between ice ages and human evolution is based on a causal connection. As selection pressure grew larger, hominids near the rim of the ice developed special abilities and invented new methods in hunting, housing, food collecting, and cooperative preparations for the future. Later, complex social structures developed and spread all over the world. The final fate of humankind is uncertain, based on the many dangerous climatic and technical developments, most of them self-made. Over the last 600 million years, 99 percent of life forms on Earth disappeared. Mankind is about 5 million years old, and Earth may be habitable for another 1 billion years.

epilogue

Our Limited Resources

Humankind depends on the Earth's resources, including hydro-carbons, coal, lignite, ores, various rocks and minerals, soil, and clean water, in order to sustain life. However, these resources, which are found in the upper few kilometers of our planet's crust, are not renewable. The development and maturation of petro-leum takes several million years (we have used most of Earth's supply in only 100 years), and even a fertile soil needs several hundred years for its maturation. Some ores, like the banded iron formations, only appear in old cratons, while hydrocarbons and coal develop in relatively young and shallow marine or swamp areas. Although groundwater, in principle, is renewable, increased consumption, especially in the fast-growing cities of third-world countries, causes tremendous shortages. While fossil resources were discussed in Chapter 11, we will now take a look at alterna-tive and sustainable energy sources and see what the future might hold in store.

Nuclear Energy

Since the mid-twentieth century, *nuclear energy* has provided another major source of power. Through the fission of radioactive elements such as uranium, electricity is generated in huge reactors. There are sources of uranium in many countries, and these reserves may ameliorate the dwindling reserves of fossil fuels. No carbon dioxide is generated by a nuclear reactor, which is a big advantage compared to the carbon dioxide emissions from coal and oil consumption. But nuclear power plants pose high security risks, and there are big problems with the transport, storage, and disposal of nuclear waste. However, the storage in a stable salt plug or a granite body is relatively safe.

Alternative Energy Sources

Many possibilities have been investigated for conserving our energy resources. These alternative energies could be able to cover a certain percentage of our increasing needs. Water power, solar energy, wind power, and biomass (such as agricultural waste) are all renewable and sustainable sources of energy.

Water Power

Countries like Brazil and Egypt cover nearly 100 percent of their electricity demand by water power. The gigantic Iguaçu power plant in southern Brazil is currently the largest hydroelectric plant in the world. A still larger power plant is under construction in China, where a massive dam will create a long lake and generate energy comparable to 12 nuclear reactors. Another hope is that the dam will prevent devastating floods. (Over the past centuries, millions of people have died or been made homeless in floods below the dam site where it is being constructed.)

A serious threat for mankind are the limited groundwater resources, especially in the subtropical and dry regions of our planet, and in the growing cities. There has always been a shortage of groundwater in arid climate zones, but today, the situation is even more threatening. Disputes over groundwater and river rights have grown from being regional to being international issues. Moreover, there is a terrifying correlation between infant mortality and contaminated groundwater in many third-world countries. Technical support from industrialized countries should be one of the most important tasks in the decades to come.

Solar Energy

The use of solar energy requires many cloudless days, which makes it a viable option for subtropical countries or countries in higher latitudes. Solar cells, mostly silicon cells, convert solar radiation into electric current. Tests in California, Spain, and the Sahara, partly supported by enormous mirrors, are promising. There are plans to use solar energy to convert hydrogen from the water of lakes and seas, then use the hydrogen in fuel cells to create electric current. Some scientists even suggest catching sunlight on Mars by huge solar mirrors and using solar cells to heat the Martian surface and atmosphere in order to prepare for future human colonies. We should keep in mind that our biomass production on Earth is also a way to use stored solar energy.

Wind Power

Wind power is another renewable energy, and countries like the United States (particularly California), Denmark, and Germany have built large "wind parks," some even at sea. The hope is that such facilities will eventually provide up to 10 percent of electricity requirements. Hybrid techniques that make joint use of solar and wind energy, with large rotors and sails, are still in the experimental stage.

Geothermal Energy

Geothermal energy originates from the accretion of the Earth—the formation of its core and the radioactive elements, mainly uranium, potassium, and thorium, in its crust. Heat is transported upward by heat conduction and some convection, while the temperature increases with greater depths. Differences in the geothermal gradient depend generally on the age of the crust. Old and cold cratons have a small gradient, while young and volcanic areas show a strong gradient, as in the neighborhood of active plate boundaries. High temperature is the first requirement for extracting geothermal energy. The second precondition is a thick, porous, and water-containing rock unit that should be—as a third condition—isolated by a non-permeable layer in the hanging wall.

Depending on the temperature and the tectonic environment, there are usually two types of plants for the generation of geothermal energy. One type uses hot *high-enthalpy* sources (often found in volcanic areas with much steam) and turbines for the production of electricity. This type of plant generates much steam and noise and is environmentally unfriendly. Temperatures have

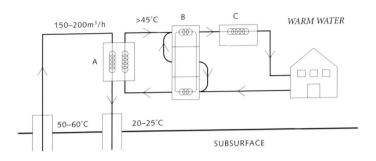

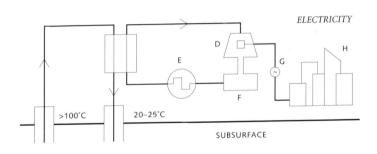

A *Heat Exchanger*	E *Pump*
B *Heat Pump*	F *Cooler*
C *Back-up Hot Water Tank*	G *Electric Current*
D *Turbine*	H *City*

FIGURE EP.1. Two types of plants for extracting geothermal energy (using so-called *doublettes*). Top: Low-enthalpy plant for rather clean heating. Bottom: High-enthalpy plant for electricity.

to be at least 100 degrees Celsius. The other type, the *low-enthalpy* sources, are used for heating of buildings, greenhouses, pools, and spas. The plant has to be in the end user's immediate neighborhood. Often, temperatures of 40 to 60 degrees Celsius are sufficient. In the sediment troughs of basins, at 2 to 3 kilometers of depth, temperatures of up to 90 degrees Celsius are often found. In the Paris basin, a large part of the city is heated by geothermal energy, and only in an exceptionally cold winter is additional heating required. The Parisian plants—like all modern plants—use a so-called *doublette*, i. e., two boreholes, one for extracting hot water from the ground and one for inducing the used (cooler) water back into the subsurface. Aggressive (calcium-, magnesium, or sulphate-containing) waters need heat exchangers. In the long run, the source rocks will get cooler, and some shortcut might take place. Therefore, this energy is not totally renewable. Still, it generally takes more than 20 years before the source rocks become unusable (and by then the plant will probably have been amortized).

Outlook

Without a doubt, all of the mentioned alternative energies will have to expand. In California, a combination of wind power, biogas, and geothermal energy satisfies the electricity needs of more than 12 percent of the population, and the exhaust of carbon dioxide has decreased by about 17 percent over the last 15 years. While mining and exploitation for fossil energy will get more and more expensive in the search for the last resources, the production cost of various alternative energies is currently decreasing.

One of the largest future hazards for humankind may be the manmade increase of atmospheric carbon dioxide. We know now that this increase is a consequence of consuming and burning coal, oil, and forests. Carbon dioxide and other trace elements absorb sunlight at certain wavelengths and contribute to the warming of the Earth's atmosphere. At present, the temperature rise is about 2 to 4 degrees Celsius per century. As a consequence, glaciers will decrease in Antarctica, Greenland, and in mountains all over the world, the temperature of the ocean water will increase, and sea level will rise. Measurements from satellites and worldwide leveling show an increase of the sea level by about 2 millimeters per year, which sums up to 2 meters in 100 years. Global warming seems to be inescapable.

Looking at the last significant warm period between ice ages, from 120,000 to 110,000 years ago, the sea level was 5 meters higher, and a part of Antarctica's ice cover had melted. Knowing that, today, about one third of mankind lives near coastlines, dramatic resettlement or the construction of giant dams may be necessary in the future. Only optimists can believe that the next ice age, long predicted by the Milankovitch Cycle, may compensate for the growing increase of temperature and sea level. Indeed, a small group of researchers thinks that small variations of the Gulf Stream, caused by warming and an inefficient sinking of cooling water near Greenland (one of the engines of the Gulf Stream), may indeed initiate the next ice age. Predictions of climate changes are indeed vague, and extended research seems mandatory.

The history of the use of natural resources is long. Hominids were able to kindle warming fires about 1.5 million years ago. Today, mankind uses about 2.5 million liters of oil per day. Five billion tons of carbon dioxide per year are blown into the air from the United States alone. In regard to the poorly understood consequences of increasing carbon dioxide levels, it is absolutely necessary to limit its input into our atmosphere. We must handle our limited resources more carefully and develop "cleaner" energy. Our belief in ever-progressing technology should not decrease our respect for Nature, the ultimate source of our existence.

Further Reading

Chapter 1: The Roots of Earth Sciences

Smolin, Lee. *The Life of the Cosmos.* New York: Oxford University Press, 1997.

Hallam, Anthony. *Great Geological Controversies.* 2nd ed. New York: Oxford University Press, 1990.

Chapter 2: The Earth in the Context of Our Solar System

Ahrens, L. H. *Distribution of the Elements in Our Planet.* New York: McGraw-Hill Book Co., 1965.

Guth, Alan H. *The Inflationary Universe: The Quest for a New Theory of Cosmic Origins.* Cambridge, Massachusetts: Perseus Publishing, 1998.

Chapter 3: The Formation of Earth and Moon

Bowring, S. A. and T. Housch. "The Earth's Early Evolution." *Science* 269 (1995): 1535.

Lucey, P. G. et al. "Topographic-compositional Units on the Moon and Early Evolution of the Lunar Crust." *Science* 266 (1994): 1855.

Chapter 4: The Interior of the Earth and the Role of Seismology

Davies, Geoffrey F. *Dynamic Earth: Plates, Plumes and Mantle Convection.* Cambridge, U.K. & New York: Cambridge University Press, 1999.

Morton, Ron L. *Music of the Earth: Volcanoes, Earthquakes, and Other Geological Wonders.* Cambridge, Massachusetts: Perseus Publishing, 1996.

Chapter 5: Rotation and Shape, Gravity and Tides

Brosche, P. and J. Sündermann (eds.). *Tidal Friction and the Earth's Rotation.* Vol. 2. New York: Springer-Verlag Inc., 1982.

Melchior, Paul J. *The Tides of the Planet Earth.* 2nd ed. New York: Pergamon Press, 1983.

Chapter 6: The Earth's Magnetic Field

Cox, Alan. *Plate Tectonics and the Geomagnetic Reversals.* New York: W. H. Freeman & Co., 1973.

Sleep, Norman H. and Kazuya Fujita. *Principles of Geophysics.* Chapters 6–7. Malden, Massachusetts: Blackwell Science Inc., 1997.

Chapter 7: Atom—Mineral—Rock

Press, Frank and Raymond Siever. *Earth.* New York: W. H. Freeman & Co., 1985.

Sorell, Charles A. and George F. Sandstrom (illus.). *Rocks and Minerals: A Guide to Field Identification.* Rev. ed. New York: St. Martin's Press, 2001.

Chapter 8: The Early Ages

Monod, Jacques. *Le hasard et la nécessité: Essai sur la philospohie naturelle de la biologie moderne.* Paris: Edition du Seuil, 1973.

Prothero, Donald R. and Robert H. Dott, Jr. *Evolution of the Earth.* 6th ed. New York: McGraw-Hill Book Co., 2001.

Chapter 9: Radioactive Dating

Dickin, Alan P. *Radiogenic Isotope Geology.* Reprint ed. Cambridge, U.K. & New York: Cambridge University Press, 1997.

Faure, Gunter. *Principles of Isotope Geology.* 2nd ed. New York: John Wiley & Sons, 1986.

Chapter 10: Plate Tectonics

Kearey, Philip and F. J. Vine. *Global Tectonics*. 2nd ed. Malden, Massachusetts: Blackwell Science Inc., 1996.

Press, Frank and Raymond Siever. *Earth*. Chapter 20. New York: W. H. Freeman & Co., 1985.

Chapter 11: The Crust of the Earth

Keary, Philip and Michael Brooks. *An Introduction to Geophysical Exploration*. Malden, Massachusetts: Blackwell Science Inc., 1991.

Meissner, Rolf. *The Continental Crust: A Geophysical Approach*. San Diego, California: Academic Press, 1986.

Phinney, Robert A. (ed.). *The History of the Earth's Crust: A Symposium*. Princeton, New Jersey: Princeton University Press, 1968.

Chapter 12: Formation of Mountains and Basins

Duba, A. G. et al. (eds.). *The Brittle-Ductile Transition in Rocks: The Heard Volume*. Geophysical Monograph, no. 56. Washington, D.C.: American Geophysical Union, 1990.

Martin, H. and F. W. Eder (eds.). *Intracontinental Fold Belts: Case Studies in the Variscan Belt of Europe and the Damara Belt in Namibia*. New York: Springer-Verlag Inc., 1983.

Chapter 13: New Discoveries, New Concepts

Campbell, I. H. and R. W. Griffiths. "Implication of Mantle Plume Structure." *Earth and Planetary Science Letters* 99 (1990): 79.

Meissner, Rolf and W. Mooney. "Weakness of the Lower Continental Crust: A Condition for Delamination, Uplift, and Escape." *Tectonophysics* 296 (1998): 47.

Chapter 14: The Phanerozoic: The Last 600 Million Years

Lane, H. Richard et al. (eds.). *Fossils and the Future: Paleontology in the 21st Century.* Kleine Senckenberg-Reihe, no. 74. Frankfurt am Main, Germany: Senckenberg, 2000.

Donavan, Stephen K. (ed.). *Mass Extinctions: Processes and Evidence*, New York: Columbia University Press, 1989.

Chapter 15: Biological Evolution

Darwin, Charles. *On the Origin of Species by Means of Natural Selection or the Preservation of Favoured Races in the Struggle for Life.* Reprint ed. New York: Bantam Books, 1999.

Caldeira, K. and J. F. Kasting. "The Life Span of the Biosphere Revisited." *Nature* 360 (1992): 721.

Epilogue: Our Limited Resources

Stobaugh, Robert B. and Daniel Yergin (eds.). *Energy Future: The Report of the Harvard Business School Energy Project.* New York: Random House, 1979.

Houghton, John. *Global Warming: The Complete Briefing.* 2nd ed. Cambridge, U.K. & New York: Cambridge University Press, 1997.

Index

Page references in italics indicate figures (f) and tables (t).